TV OR CATV?
A STRUGGLE FOR POWER

Edward V. Dolan

NATIONAL UNIVERSITY PUBLICATIONS
ASSOCIATED FACULTY PRESS, INC.
Port Washington, N.Y. • New York City • London

Manufactured in the United States of America

Published by
Associated Faculty Press, Inc.
Port Washington, N.Y.

Library of Congress Cataloging in Publication Data

Dolan, Edward V., 1926–
  TV or CATV?

  (National university publications)
  Bibliography: p.
  Includes index.
  1. Cable television—United States. 2. Television
broadcasting—United States. I. Title. II. Title:
T.V. or C.A.T.V.?
HE8700.72.U6D64 1984   384.55'56'0973   83-12264
ISBN 0-8046-9329-3

Contents

ACKNOWLEDGEMENTS

I wish to express my thanks to all those who were of assistance to me in writing this book. I particularly want to note the contribution of the cable-communications directors who participated in the major market cable survey (see the Appendix). Of particular aid were Richard A. Borten, principal consultant on cable for the city of Boston, and Frank V. Greif, former director of cable communications for Seattle, Washington, and past president of the National Association of Telecommunications Officers and Advisors. I also want to thank Janet Cratsley of the National League of Cities and Sara L. Engelhardt of the Carnegie Corporation.

I am deeply grateful for the help provided by the staffs of the following libraries: the Boston Public Library, the Boston University Library, the Emerson College Library, The Quincy Public Library, the Holmes Public Library, and most especially, the Brockton Public Library.

ABOUT THE AUTHOR

As an attorney practicing in Massachusetts, Edward V. Dolan has had an acute interest in communications law for many years. This book is an outgrowth of that interest. He is a member of the Massachusetts and Federal Bar Associations and received his LL.B. from the New England School of Law in 1950.

PREFACE

When radio emerged as a mass-communications medium, the Congress of the United States was presented with a problem. The airways were not private property; they belonged to the people. Because the electromagnetic spectrum had a finite capacity, it could not be used by everyone. A system had to be devised that offered access to some — broadcasters — and excluded everyone else.

Despite early legislative efforts (the Radio Act of 1912), the public airways were in a state of electronic anarchy, a "simply intolerable" condition in the words of Herbert Hoover, Secretary of Commerce and Labor. Radio stations increased power and changed frequencies at will. The broadcasting industry demanded regulation by the federal government — something of a departure from their current view.

Congress responded by passing the Radio Act of 1927, followed by the present law, the Federal Communications Act of 1934. Under both statutes those using the electromagnetic spectrum were licensed and designated "public trustees", charged with the duty of broadcasting in the public interest.

It was a unique solution, one that made considerable demands on human nature. A businessman, ordinarily devoted to maximizing profits, had now to assume another role — that of public trustee. It was a demonstrable conflict of interests that the United States Congress could hardly ignore. The solution was to create a regulatory commission — first the Federal Radio Commission then the Federal Communications Commission — to oversee the broadcasting industry and insure that the public interest was protected. The regulators were to act as watchdogs and policemen for the American people.

The business of commercial broadcasting has no counterpart in our free-enterprise economy: a government-authorized, moneymaking business using public property without compensation in a protected market with no limitation on profits. For the select few, the enabling legislation could not have been more favorable had they written it themselves.

These arrangements came to pass long before public-interest advocacy became fashionable, before most people fully appreciated the implications. For most Americans, radio was a miracle that brought entertainment into their homes; few worried about who controlled it.

Some of the more astute observers understood what was happening and foresaw its consequences, but they were powerless to restrain the already powerful broadcasting lobby. Politicians found themselves in an exposed position, fearing the power of radio as an instrument of retribution. Opposition to the industry's wishes was considered dangerous politically.

With the advent of television, broadcasters' power increased enormously. Television has become the most potent instrument of political power since the invention of the printing press and those who control it are the most powerful people in the country.

From a political perspective, it would be naive to expect fundamental changes in broadcasting. The symbiotic relationship that exists between legislator and broadcaster, sustained by implicit blackmail, has protected the industry and its vital interests for more than two generations.

Fortunately, not all change is dependent on the workings of the political process; technology also plays a part. We are now in the early stages of a new communications revolution, the third in this century. It is commonly called cable television — a misnomer. A more accurate term would be *cable communications,* a multipurpose communications system as distinguished from a medium, i.e., radio or broadcast television. Cable has the inherent technical capacity to be an interactive communications system in the audio, video and digital modes. Its channel capacity — unlike radio and broadcast television — can be expanded to meet demand. Because of cable's enhanced communication's capacity, cablecasting holds the promise of democratic participation, a system that can accommodate those seeking to use it — from program supplier to average citizen. Cable can become a true marketplace of ideas with access to its channels a *matter of right,* something no other medium, whether print or electronic, offers.

Not everyone shares this vision of what cable might be. Cable interests, understandably, seek maximum control, the kind of control that commercial television enjoys. Some demand print media rights. They resent the franchising process and the concessions most cities have demanded in return for the use of public property. Intent on invalidating provisions in the franchises they have freely negotiated, cable operators are now doing what broadcasters have done for decades: lobbying Congress for favorable legislation that will fundamentally alter local authority over cable. This ongoing power struggle deserves our attention if cable is not to go the way of broadcast television

1

Toiling in the Public Interest

In what other business can a moderately astute operator hope to realize a 100% a year on tangible assets or lay out $150. for a franchise and in a few years time he can peddle it for 50 million.
— *The Alfred I. duPont–Columbia University Survey of Broadcast Journalism, 1968-1969*

The expression "a license to print money" when applied to a television license is an exaggeration but one holding a large measure of truth. A government-issued license to operate a television station, particularly a VHF (very high frequency) license in a major urban market having a network affiliation, is an extremely valuable asset for which the chosen few pay nothing. It is a gift. The recipient promises only to serve as a public trustee and broadcast over the airways "in the public interest." The licensee uses the assigned frequency; he does not own it. The airways — i.e., the electromagnetic spectrum belong to all the people.

Though acquiring no property right per se, the licensee can sell his right to use the public airways on the open market to the highest bidder, enriching himself in the process. The government and the people get nothing except a new trustee.

The government limits the number of licenses, thereby creating a shared-monopoly environment conducive to high profitability. As the population grows within a market, the broadcaster's circulation expands, thus making "his" air time more valuable.

Needless to say, there has never been a shortage of would-be public trustees eager to serve the people. This was especially true in the late 1940s and early 1950s when most of the valuable television licenses were granted. Getting in on this extraordinary government largesse meant getting in early. Perceptive members of the communications establishment understood the considerable potential of this new medium — the spectacular success of radio a generation earlier serving as a model. Consequently, the number of applicants far exceeded the number of licenses to be awarded, precipitating a ferocious lobbying campaign. Every power broker in Washington had a favorite candidate for public trusteeship.

By the late fifties most of the valuable licenses had been granted; what remained were UHF licenses. With a few exceptions, UHF stations did not have a network affiliation and had little to offer the viewer except recycled game shows, sit-coms, and old, old movies.

The rationale sustaining our system of public broadcasting has no precedent. When government sells or leases its assets — that is to say, our assets — it demands and gets fair market value. Favored treatment, let alone an outright giveaway, would cause a national scandal. Even though the rights granted to broadcasters are intangible, thus distinguishable from other government assets, the benefits conferred have monetary value. In fact, the intangible right to use the public airways is more valuable than an oil or mineral lease. Eventually these assets will be depleted; a television license can last forever despite the largely pro-forma renewal process.

The quid pro quo for a television broadcasting license is a promise by the public trustee to serve the pulbic interest. This nebulous standard is defined by the licensee, who alone decides what the public will see and hear. Subject only to the Fairness Doctrine and the equal-opportunities requirements of the Federal Communications Act, the licensee had absolute discretion on what will be or will not be broadcast.

Given the free use of public property, in a protected market that assures large profits, with near-absolute discretion in programming, one might reasonably expect good faith on the part of our public trustees, that they would take seriously their promise to serve the public interest. Unfortunately such an entirely reasonable expectation has not been fulfilled. In retrospect, this result could and should have been anticipated. The theory of public trusteeship within the context of commercial broadcasting is inherently contradictory. The motivation of a businessman-broadcaster is to maximize profits for his stockholders; the motivation of a public trustee is to serve the interest of the general public. These dual motivations put the public trustee-businessman in an impossible situation. If he places the public interest ahead of maximum profits, he may find himself in trouble with his stockholders; if he serves the stockholders' interests at the expense of the public interest, he defaults on his obligations as a public trustee. At best, the rationale for public trusteeship is naive; at worst, it is cynical and fraudulent.

Broadcasters argue that there really is not a conflict of interest, that what is good for the business of broadcasting is also good for the public interest, a proposition that enjoys little support from objective observers. The reality is that the concept of public trusteeship within the context of commercial broadcasting represents little more than a fig leaf, a designed obfuscation.

Since coming to maturity in the 1950s, television has enjoyed a remarkable success. It has become immensely profitable, beyond even the most optimistic projections of its investors. While most businesses are amortized over a period of years, television stations with network affiliations in major markets frequently recover their invested capital in a matter of months. The American love affair with television has created an economic bonanza for the chosen few, one that continues to flourish.

The three networks and their wholly owned and operated television stations have had a record of spectacular earnings. An early congressional investigation of the industry is highly illuminating. Chairman Emanuel Celler invited FCC Chairman George C. McConnaughey to testify before the Antitrust Subcommittee of the Judiciary Committee of the House of Representatives, Eighty-Fourth Congress, Second Session. The following is an excerpt.[1]

The Chairman: First with respect to revenues, it is correct, is it not that for 1955 Columbia Broadcasting System and the National Broadcasting Company networks and their nine wholly owned television stations had broadcast revenues of $312,658,470?
Mr. McConnaughey: Yes, sir.
The Chairman: That is correct?
Mr. McConnaughey: Yes, sir; that is correct.
The Chairman: It is also correct, is it not, that these revenues constituted 41.99 per cent of the revenue for the entire television industry?
Mr. McConnaughey: That is correct, sir.
The Chairman: It is also correct, I take it, that the revenue of CBS plus its 4 television stations excluding radio was $153,614,317 and was equal to 20.63 per cent of the revenue of the entire television industry?
Mr. McConnaughey: That is correct, Mr. Chairman.
The Chairman: It is also correct that CBS revenue from network operations in 1955 was $121,953,917, and NBC revenue from network operations was $124,353,526?
Mr. McConnaughey: That is correct, sir.
The Chairman: And it is correct that the combined revenue from CBS' and NBC's network operations in 1955 was 87.2 percent of the broadcast revenue of all television networks?
Mr. McConnaughey: That is correct..
The Chairman: Now turning to net income before federal income tax of CBS and NBC operations, it is correct, I take it, that in 1955 the CBS and NBC networks together with their 9 stations had net income before taxes of $65,050,186?
Mr. McConnaughey: That is correct, sir.
The Chairman: And that this is 43.4 per cent of the net income before taxes of the entire television industry?
Mr. McConnaughey: Yes, sir.
The Chairman: That CBS and its 4 stations had a net income of $34,870,837 or 23.2 per cent of the income of the entire industry?
Mr. McConnaughey: That is correct, sir.
The Chairman: And that NBC with 5 stations had a net income of

$30,179,349 or 20.1 per cent of the income of the entire industry?

Mr. McConnaughey: Yes, sir.

The Chairman: Now I should like to ask you some questions concerning ratio of 1955 income to total investment in broadcast property as of December 31, 1954. First, it is true, is it not, that as of December 31, 1954 CBS and NBC plus their 9 owned stations, had an investment in broadcast property of $50,067,737?

Mr. McConnaughey: That is correct, sir.

The Chairman: That is correct?

Mr. McConnaughey: Yes sir; it is.

The Chairman: All of these figures are taken from your record?

Mr. McConnaughey: Yes; they are.

Mr. McConnaughey: That is correct, sir.

The Chairman: That is correct?

Mr. McConnaughey: Yes sir; it is.

The Chairman: All of these figures are taken from your record?

Mr. McConnaughey: Yes; they are.

The Chairman: This is a reaffirmation. You have already testified, have you not, that in 1955 CBS and NBC, plus their 9 owned stations, had a net income before taxes or $65,050,186?

Mr. McConnaughey: Yes, sir.

The Chairman: This means, does it not, that in 1955 CBS and NBC recovered, both these systems recovered, 131 per cent of their total investment in broadcast property?

Mr. McConnaughey: Yes, sir, that is correct.

The Chairman: That is, in 1 year, in 1955, they recovered back 131 per cent of their total investment in broadcast property?

Mr. McConnaughey: That is correct, Mr. Chairman.

The Chairman: It is also correct, is it not, that in 1954, CBS and NBC recovered 99 percent of their total investment in broadcast property? I refer to the Bricker report, appendix chart 1; that is correct, isn't it?

Mr. McConnaughey: Yes, sir; it is correct.

The Chairman: And it is correct that CBS and NBC in 1953 recovered 53 per cent of their total investment in broadcast properties?

Mr. McConnaughey: Yes, sir.

The Chairman: Now I should like to direct your attention to the rate of return of CBS on its investment.

It is correct, is it not, that in 1955 CBS, plus its 4 television stations, reported a net income of $34,870,837?

Mr. McConnaughey: Yes, sir; that is correct.

The Chairman: And that, as of December 31, 1954, CBS had a total investment in broadcast property of $26,958,279?

Mr. McConnaughey: That is correct, sir

The Chairman: So therefore, in 1955 CBS in its network and station operations earned 129 per cent of its total investment?

Mr. McConnaughey: That is correct.

The Chairman: In 1 year it got back 129 per cent of its total investment and that year was 1955?

Mr. McConnaughey: That is in tangible broadcast property; yes.

The Chairman: That is what I mean, tangible broadcast property. In 1954 CBS and its television stations had a return of 108 per cent of its total net investment; is that right?

Mr. McConnaughey: That is correct.

The Chairman: It is also correct that in 1953 CBS earned 54 per cent return on its net investment?.

Mr. McConnaughey: That is correct, sir.

The Chairman: Now turning to NBC, it is correct, I take it, that in 1955 NBC, plus its 5 television stations, had a net income before taxes of $30,179,349?

Mr. McConnaughey: That is correct.

The Chairman: It is also a fact that as of December 31, 1954, NBC had a total investment in broadcast property of $23,109,458?

Mr. McConnaughey: That is correct.

The Chairman: So that in 1955, this same year, NBC and its owned stations had a return before federal income taxes of 133 per cent on its investment

Mr. McConnaughey: That is correct.

The Chairman: And in 1953 that same company had a return of 52 per cent of its investment?.

Mr. McConnaughey: That is correct, sir.

The Chairman: Now I want to ask you some questions on investment of network-owned television stations, stations owned by CBS and NBC. First, it is correct, is it not, that the nine stations owned by CBS and NBC earned an income in 1955, before taxes, of $30,081,992?

Mr. McConnaughey: That is correct, sir.

The Chairman: And that as of December 31, 1954, CBS and NBC had an investment in their stations of $9,973,056?

Mr. McConnaughey: That is correct.

The Chairman: So, therefore, it is correct, is it not, that in 1955 the television stations owned by CBS and NBC earned 307 per cent of their total investment in broadcast property?

Mr. McConnaughey: That is right, sir.

The Chairman: It is correct, too, is it not, that in 1954 the TV stations owned by those 2 companies earned 330 per cent on their investment?

Mr. McConnaughey: That is correct.

The Chairman: And in 1953 those same companies earned 239 per cent of their investment?

Mr. McConnaughey: That is correct, sir.

The Chairman: Turning to the return on investment of television stations owned by CBS, it is correct, is it not, that in 1955, the same year, the 4 television stations owned by CBS earned 282 per cent on investment?

Mr. McConnaughey: That is correct, sir.

The Chairman: It is also true that in 1954 stations owned by CBS earned 370 per cent on investment?

Mr. McConnaughey: That is correct.

The Chairman: And in 1953 the return was 226 per cent on investment?

Mr. McConnaughey: That is correct, sir.

The Chairman: Now as to NBC, in 1955 NBC's 5 television stations earned 335 per cent on investment?

Mr. McConnaughey: That is correct, sir.

The Chairman: And that in 1954 television stations of NBC yielded 297 per cent on investment?

Mr. McConnaughey: That is right, sir.

The Chairman: And in 1953, 251 per cent on investment?

Mr. McConnaughey: That is correct, sir.

The Chairman: Now at this point I offer in evidence charts showing 1955, 1954, and 1953 broadcast revenues, expenses, and income of the entire television network industry, broken down by television networks, television-network-owned stations, and television networks plus their 16 network-owned stations.

Mr. McConnaughey: You refer of course, when you say investment to tangible broadcast properties?

The Chairman: That is right; I limit myself to that, naturally, which is their investment of course.

Mr. McConnaughey: That is correct.

The Chairman: That does not take into consideration whatever investment they have, if any, in talent and things of that sort. We are speaking of net returns; and, in those net returns, expenses for talent and other costs of operation are considered?

Mr. McConnaughey: That is right, sir.

The Chairman: I would like to ask you about financial data with regard to the network-owned stations. You have made available to this committee at its request, certain financial data on network-owned television stations.

Now in 1954 CBS owned three television stations, namely, KNXT in Los Angeles, WBBM in Chicago, WCBS in New York. That is correct, is it not?.

Mr. McConnaughey: Yes, sir.

The Chairman: The net income of these three stations in 1954 before federal income taxes was $12,276,443?

Mr. McConnaughey: Yes, sir.

The Chairman: The net investment in tangible broadcast property of these three stations as of December 31, 1953, was $3,332,023; is that correct?

Mr. McConnaughey: Yes, sir.

The Chairman: This means, does it not, that in 1954 the 3 CBS television stations recovered 370 per cent of their total investment in broadcast properties?

Mr. McConnaughey: That is correct, sir.

The Chairman: In 1955 CBS owned four television stations, KNXT, Los Angeles, WBBM, Chicago, WXIX, Milwaukee, and WCBS, New York. That is correct, isn't it?

Mr. McConnaughey: Yes, sir.

The Chairman: The net income of these four stations in 1955 before federal income taxes was $14,505,459?

Mr. McConnaughey: That is right.

The Chairman: The net investment in tangible broadcast property of those 4 stations as of December 31, 1954, was $5,146,981?

Mr. McConnaughey: Yes, sir.

The Chairman: This means, does it not, that in 1955 the 4 CBS television stations recovered 282 per cent of their total investment in broadcast property in that one year?

Mr. McConnaughey: Yes, sir; that is right.

The Chairman: Let me ask you about just one CBS television station particularly; namely, WCBS in New York.

Isn't it a fact that WCBS in New York had a net income before federal income taxes in 1953 or $5,571,777?

Mr. McConnaughey: That is correct, sir.

The Chairman: And that WCBS, the same station, had a net investment in tangible broadcast property as of December 31, 1952, of $528,911?

Mr. McConnaughey: That is right, sir.

The Chairman: Which means, does it not, that WCBS recovered 1,053 per cent of its total investment in broadcast property in 1953?

Mr. McConnaughey: That is correct, sir.

The Chairman: In 1954 WCBS had a net income of $8,206,416 before federal income taxes?

Mr. McConnaughey: Yes, sir.

The Chairman: And that WCBS had a net investment in tangible broadcast property as of December 1, 1953 of $447,420?

Mr. McConnaughey: That is correct, sir.

The Chairman: This means, does it not, that in 1954 WCBS made 1,834 per cent on its total investment in broadcast property?

Mr. McConnaughey: That is correct, sir.

The Chairman: In 1955, WCBS had a net income before federal income taxes of $9,375,339?.

Mr. McConnaughey: Yes, sir.

The Chairman: And that WCBS had a net investment in tangible broadcast properties as of December 31, 1954 of $409,484?

Mr. McConnaughey: That is correct, sir.

The Chairman: This means, does it not, that in 1955, and I give emphasis to these figures, WCBS recovered 2,290 per cent on its total investment in broadcast property?

Mr. McConnaughey: That is correct, sir.

The Chairman: You would say, would you not, those are high profits?

Mr. McConnaughey: Extremely high profits.

This extraordinary record of profitability in the television industry is a direct consequence of governmental involvement in the granting of valuable broadcasting licenses to the favored few, the shared-monopoly environment, the absence of any limitations on profits.

Given the nature of the industry, increasing profits are virtually assured, the reason being that revenue and profits bear little relationship to operating costs. Once built, a television station can transmit its signal to an ever-expanding audience without incurring proportional costs — a distinct advantage over most other business enterprises. Television is unique; there is no business quite like it.

With only occasional dips, the curve line on profits heads upward, insulated from the economic woes that affect other businesses. Television is a seller's market and seems destined to remain so, at least in the near term, until cable comes to maturity offering some badly needed competition.

Although the industry as a whole is extremely profitable, not all television stations share in the wealth. Television stations fall into three general categories: major market, network-affiliated, VHF stations making exceedingly high profits; independent VHF stations in the larger markets and network-affiliated VHF stations in medium and smaller markets making ex-

cellent profits; and nonaffiliated UHF stations making modest profits or losing money. In the latter category, some UHF stations have become profitable by specializing in sports and movies.

The television business is a numbers game. Broadcasters are brokers who
sell an audience to advertisers. Unless a great number of people watch a program, advertisers are ordinarily not interested. Thus, stations in the VHF
bandwidth with a network affiliation — the stations most people watch — do
well. But just how well is not certain. Broadcasters and the commercial networks tend to be reticent about their profits, and not without cause. If the
public knew how much money was being made by those licensed to use the
spectrum, some serious questions might arise.

The FCC, to which broadcasters must report annually, refuses to disclose
profits of individual licensees, opting instead on a market reporting system.
By lumping the profits of all stations in a given market, the exact profits of
any one can be concealed. Apparently the FCC holds on to the belief that
what the public trustees of our airways make in the discharge of their public
trust is none of the public's business. Since competing stations in the same
market know what other stations are charging for their time, profits can be
readily calculated. Only the public that broadcasters purport to serve is
uninformed.

Despite the industry-inspired, bureaucratic dissembling by the FCC, a
careful review of the published data can be revealing. In some of the larger
markets where marginally profitable or money-losing UHF stations are
lumped with highly profitable, network-affiliated VHF stations or network-
owned stations, one is obliged to speculate — which is, of course, the intent
of market reporting. However, in the medium-sized markets having no independent, nonaffiliated stations, a clearer picture emerges.*

Columbus, Ohio, has three stations, all VHF, all network-affiliated. In
1980, the three stations reported net broadcast revenues of $37,234,512 and
income of $14,028,482.

Birmingham, Alabama, has three stations, each with a network affiliation
— one on the UHF bandwidth. In 1980, the total net broadcast revenue was
$26,171,260 with broadcast income of $12,453,397.

The Toledo, Ohio, market with three stations — one, UHF — reported
broadcast revenues of $19,787,031 and broadcast income of $6,130,659.

Profit performances like these in medium-sized markets are not unusual
for network-affiliated stations. But in the Dallas-Fort Worth and Houston-

---

*The FCC does not report on markets with fewer than three reporting stations. All statistics
quoted are taken from *Television Factbook,* No. 50, 1981-82, pp. 69a-70a.

Galveston markets, profits of only 30 or 40 percent could get a station manager fired.

In Dallas-Fort Worth there are five stations: three VHF network-affiliated stations, one independent VHF station, and one independent UHF station — KXTX-TV, Channel 39 — licensed to the Christian Broadcasting Network, Inc. Total revenue for the five stations was $117,835,499; total pretax profits were $62,017,163. The *average* for each station was $12,403,432. With two nonaffiliated stations reporting, the average is quite deceptive. Clearly the three network-affiliated stations exceeded the average profit of over 50 percent by a considerable margin.

A similar success story exists in the Houston-Galveston market with five stations: three network-affiliated VHF stations and two independent UHF stations. Total broadcast revenue for the market was $108,386,612 with an income of $58,870,038. Once again the average of $11,774,007 is grossly distorted by the inclusion of two independent, nonaffiliated UHF stations.

Inspirational though the Texas markets may be for profit-maximizers, they are not the most rewarding examples of private-enterprise broadcasting. For the truly profitable stations, one must look to the largest markets: New York, Los Angeles, and Chicago. Here each of the commercial networks has an owned-and-operated station, a monopoly on network programming, and huge audiences, elements which in television guarantee high profitability.

2

GUARDIAN AT THE GATE:
THE FEDERAL COMMUNICATIONS COMMISSION

The question is one of intent: What did Congress really have in mind when it created the Federal Communications Commission? Was it designed to protect the broad public interest as its proponents claimed, or was it merely a skillful legislative exercise in legerdemain, locking into place an already powerful commercial broadcasting industry? Congress was obliged to deal with the issue of licensing broadcasters; someone had to select the favored few. Congress itself could not do it — at least, not directly, that would have caused a legislative circus. So an institution had to be created to do the job, an "independent" regulatory agency known as the FEDERAL COMMUNICATIONS COMMISSION.

The designation of a federal agency as independent should not be interpreted literally; Congress never delegates all its powers free of constraints, express or implied. Clearly the FCC was to exercise its powers subject to strong congressional and executive oversight. It was not to have judicial insulation, the kind of independence that removed it from politics. The FCC's real independence can be placed in perspective when one examines the levers of power: the President appoints all commission members subject to Senate approval; the Office of Management and Budget approves the commission's budget; Congress appropriates the money. In short, there are many ways to reward or punish, and thus to exercise control. The FCC was to have power, but only within the parameters of existing power, a legislative creation that appeared to be something that it was not.

The statutory grant of authority is exceptionally broad with the FCC exercising quasi-legislative powers in rulemaking, such rules having the force of law; as prosecutor in enforcing these rules and other applicable law; and as judge and finder of facts. It can levy fines, even revoke licenses. On paper, these powers are impressive, even excessive. Congress, it would seem, had created the tools to do the job; if it was not properly done, it was the FCC's fault.

This was a politically astute resolution of a highly sensitive and explosive problem, one with which Congress had no choice except to deal with. By delegating power to the FCC, and designating it an "independent" regulatory agency, Congress created a lightning rod, one that offered insulation from political storms. Those having a complaint against broadcasting or broadcasters were obliged to take it to the FCC. This regulatory solution to matters

11

which Congress cannot or chooses not to handle itself is now a standard approach. Unfortunately the solution is seldom either effective or permanent; there is an entirely predictable cycle: A regulatory agency is created; the regulated industry moves in and captures the regulators; the agency fails in its assigned mission, and there are demands for change; when the pressure builds, Congress responds by enacting amendments, by reorganizing the agency, or abolishing it and creating a new independent agency — and so it goes.

In this century, Congress has enacted three laws dealing with electronic mass communication, with a fourth under study. It is clear that in enacting legislation regulating the use of the public airways, the members of Congress were neither ignorant nor naive concerning the substantive issues involved; politicians are acutely sensitive to and interested in matters concerning the mass media. The current law, the Communications Act of 1934 — was enacted less than eight years after the passage of the Radio Act of 1927, legislation which had created a record of pervasive corruption and political influence-peddling by members of Congress in behalf of favored constituents. Congress failed to reform the system because it had long since come under the influence of the broadcasting lobby.

All this wheeling and dealing came into play by virtue of the marriage of technology to commerce. While dots and dashes could deliver a message, only the human voice could sell merchandise to a mass audience. Radio in the early twenties had come of age; it was no longer a toy; it could do useful things, like making money. Entrepreneurs by the hundreds entered this new and exciting field, and in so doing created a state of electronic anarchy. There were too many free spirits about, jumping frequency and increasing power at will. The need for regulation was obvious.

At the first Washington Radio Conference in 1922, Secretary of Commerce and Labor, Herbert Hoover asked the broadcasting industry for guidance as to how he should administer the ineffective Radio Act of 1912: "It is the purpose of this conference to inquire into the critical situation that has now arisen through the astonishing development of the wireless telephone; *to advise the Department of Commerce as to the application of its present powers of regulation,* and further to formulate such recommendations to Congress as to the legislation necessary" (emphasis added).[1]

Represented at the conference was a broad cross section of broadcasting and manufacturing interests, including the Radio Corporation of America, General Electric, Westinghouse, and the American Telephone and Telegraph Company. Nobody represented the interests of the consumer; at that time such a thought would have been dismissed as absurd. Those who understood the potential of this new medium wanted it structured to meet their interests, not those of the broader public interest. Even before the Radio Act of 1927, politicians who might have represented the rights of citizens were under extreme pressure from radio-station owners and manufacturing interests. To oppose those powerful interests made little sense, politically speaking. There was no demand that Congress "do something" about radio; the American public was too preoccupied by the pleasure, excitement, and benefits of the new medium to be concerned about long-term policy considerations. People were more than satisfied with this electronic miracle; best of all, it came into the home "free," so the price was right.

In such an environment, it is small wonder that the Radio Act and the Communications Act faithfully represented the interests of the broadcasting establishment. We should not be either shocked or surprised by this result; Congress always bends to the prevailing winds. Though the political landscape has changed in the past four or five decades, we remain the hostages of precedent, of entrenched institutional power. The FCC, created in that unenlightened period, can hardly be an instrument for relief. It is like Congress a coopted institution, one that must adapt itself to existing power relationships; it is not independent, nor can it become so by congressional declaration. Something close to judicial insulation would have had to be created if it was to have any chance of success. Neither Congress nor the broadcasting lobby wanted such an arrangement. What they wanted was the appearance of independence and the reality of control, and this is precisely what they got.

The selection process for FCC commissioners is an integral part of the control mechanism. A prospective commissioner undergoes careful scrutiny by the President, members of the communications committees of Congress, and the ever-watchful broadcasting industry. A prospective candidate's public statements and private attitudes are weighed and evaluated; interested parties are frequently consulted prior to nomination. This screening process usually eliminates anyone openly hostile to vested interests. A commissioner who might make life unpleasant or difficult.* The selection of the "right" commissioner is, of course, important; it eliminates the necessity to cajole or threaten. He knows what is and is not expected of him; he knows where the power is and acts accordingly. If he conducts himself properly, he may be rewarded by reappointment, or if he chooses, employment in the industry he formerly regulated. Several have become network vice-presidents; many have left to become members of law firms practicing before the commission. Some three out of four ex-commissioners, according to a Ralph Nader survey, leave to accept employment, or in the case of attorneys, to become retained by the communications industry, a decidedly unhealthy state of affairs that can only lead to the public interest being compromised.

Professor Bernard Schwartz, an expert on administrative law, was retained to examine the workings of government regulatory agencies in the

---

*From time to time the process breaks down, and a truly independent candidate is appointed. Two examples are Newton Minow and Nicholas Johnson.

1950s. This was the time when many television licenses were being awarded. In examining the licensing process, he uncovered evidence of secret dealings between FCC commissioners and applicants including gifts, paid vacations, and a "loan" to one commissioner from a successful applicant. Another commissioner obtained expenses for a trip he made from three sources: a television station, the National Association of Radio and Television Broadcasters, and the United States Treasury. The same commissioner later served as chairman of the FCC.

One of the more troubling findings by Professor Schwartz was the lack of consistently applied standards in comparative hearings, instances where more than one legally qualified applicant sought a license. The FCC had established three criteria: local ownership, broadcasting experience, and integration of ownership. The fair and impartial application of these criteria was important since nearly every candidate met the minimum legal requirements for getting a license. After examining sixty cases, the professor found "a most disturbing inconsistency on the part of the commission in applying its criteria." [2]

Having created guidelines, the FCC proceeded to ignore them whenever it suited its purposes. "The commission," Professor Schwartz declared, "juggles its criteria in particular cases so as to reach almost any decision it wishes and then orders its staff to draw up reasons to support the decision." [3]

Judge James M. Landis in a scathing report to President John F. Kennedy, came to the same conclusion: "The anonymous opinion writers for the commission pick from a collection of standards those that will support whatsoever decision the commission chooses to make." [4]

In the course of his review, Professor Schwartz developed information on seventeen license grants warranting a full-scale inquiry — matters such as ex parte intervention, improper political influence and favors. Hearings were held by the congressional subcommittee on just two cases; the others were quietly dropped.

A short time later, Professor Schwartz was summarily fired. This harmless academic had become an investigatory tiger; he was off his leash, conducting a "runaway" investigation. This conclusion became painfully evident when the professor directed his attention to the personal involvement of a member of Congress in questionable activity — and not just any member, but one Oren Harris, chairman of the House Interstate and Foreign Commerce Committee, the committee with jurisdiction over the FCC.

Congressman Harris had become part owner of television station KRBB shortly after becoming committee chairman, acquiring a 25 percent interest for the bargain price of $500 dollars in cash and a promissory note for $4,500. Chairman Harris's ownership had a marked influence on the fortunes of KRBB. An increase in power was granted without a hearing. A previous application had been denied on the grounds that the station was not in a sound financial condition, an impediment quickly corrected when the Radio Corporation of America extended more than $200,000 in credit and a bank loaned an additional $400,000.

Having become aware of these facts, Professor Schwartz refused to ignore them and in so doing, he embarrassed a senior member of the congressional club. He thereby broke all the rules; staff employees were expected to exercise more discretion.

A much more interesting example of how to convert power into money is the Lyndon B. Johnson success story. Despite drawing a government paycheck all his adult life, he became a millionaire several times over. How? By assuming the burdens of public trusteeship. The LBJ Company applied for and was granted a television license in Austin, Texas, during the time that Johnson was Senate majority leader, a position of power no administration could afford to ignore. Not unexpectedly, the FCC concluded that the LBJ Company would make a suitable public trustee. In granting this license, the FCC conferred a monopoly on the LBJ Company, one that was maintained for thirteen years in the Austin market.[5]

Although power grabs by elected officials are odious, theya re not illegal. The law does not prevent a congressman or senator from acquiring a license to operate a radio or television station while in office despite the clear advantage it gives an incumbent. The commission could eliminate many such abuses. It has the power; what it lacks is the will.

A chilling example of the commission's disregard of the public interest was the proposed sale of the American Broadcasting Company to ITT. This giant multinational corporation, famous for its foreign intrigues, wanted to acquire ABC. Its motive was, it claimed, entirely benign; ABC was simply a good investment. Some suspected other intentions, a desire to co-opt a major communications medium along with its news and public-affairs departments. Four of the seven commission members saw nothing wrong this plan and voted to approve the sale. Only the time intervention of the Department of Justice blocked it.

Control of the FCC by the broadcasting lobby has been and remains its first priority. A truly independent commission would pose a threat to the industry's vital interests by insisting on high public-interest standards and enforcing its rules by denying license renewals. The commission's broad statutory powers are seldom invoked, the result being that broadcasters operate as they please, which is to make as much money as they can.

When license-renewal time arrives, broadcasters might be expected to defend their stewardship as public trustees, which would necessarily involve the kind of progressming they offered to the viewing public. As guardian of the public interest, one might reasonably expect the FCC to examine what the licensee had done in the past to justify renewal. Broadcasters bitterly protest such a review claiming this would be censorship. As licensees they have sole discretion concerning programming; it is none of the commission's business.

Many commissioners accept this argument. Others, such as Nicholas Johnson, contend that what broadcasters do in their programming is relevant to the licensing-renewal process. If the FCC cannot examine a licensee's past performance, what precisely is its function in the renewal process? Under this proscription a licensee would be free to broadcast twenty-four hours of cartoons — or commercials — and never worry about losing his license.

Given the political realities, it is unrealistic to expect the commission to act as a vigorous advocate of the public interest. It is not truly independent. Nor would it be realistic to expect action by Congress. Like its creation, the Federal Communications Commission, Congress yields to the imperatives of power.

3

Congress: Public Interest versus Private Interests

You learn that a congressman is under terrific pressure from his local broadcaster. Chances are that the broadcaster "gives" him some time on the air for a program to report to his constituents. The broadcaster may or may not support his campaign for reelection, and the broadcaster may own the local newspaper as well — and if this is the case, there is probably not another constituent in his district who means more to the congressman, and that may even include his wife!
— Newton N. Minow, former FCC Commissioner, 1965

Politics is a highly competitive activity, perhaps the most competitive activity in a highly competitive society. Winning is everything, the measure of success, the key to power and influence. When one chooses a political career, he or she is or must soon become a realist, pragmatically working the levers of power. To ignore political realities is to render oneself ineffective.

One of these realities is the power of the mass media, its influence on the political process. The larger the constituency, the more important it becomes to the candidae or officeholder. Getting the message out to the voters is central to every political campaign; failure to communicate effectively generally results in defeat. Reaching voters in a large congressional district on a person-to-person basis is an exceedingly difficult task; it is virtually impossible in a statewide campaign. Thus the role played by the mass media, and especially the electronic mass media, can make the difference between victory and defeat. The support of the mass media is a candidate's most valuable asset. Even the decision to run for office may be influenced by the prospect of favorable coverage.

This political environment predated the development of the electronic mass media. Newspapers have made and broken political careers for generations, and nobody with any regard for the First Amendment would try to restrict this power. Freedom of the press, the right to free expression, is the cornerstone of a democratic society.

Publishing is a right; broadcasting is a privilege. Newspapers use private property; broadcasters use public property. Newspapers need no governmental license to publish; broadcasters must obtain a license since only a limited number can use the electromagnetic spectrum.

Because of its special character, television has a duty to serve the public interest. Newspapers, as instruments of private power, have no such obligation. They are free to advance private interests over the public interest — and

17

frequently do so. They are free to be biased or objective; fair or unfair, accurate or inaccurate. Freedom of the press is a publisher's right; it has nothing to do with the reader or the person being written about. If one is the victim of unfair or untruthful press coverage, you can ask for a correction or demand a retraction. The publisher may honor your request if he chooses, but he has no duty to correct his mistakes. Nobody can make a newspaper publish anything; neither can you buy space unless the newspaper wants to sell it. The only remedy is a suit for libel which is costly, time-consuming, and difficult to prove.

Some scholars of constitutional law suggest that a narrow, carefully drafted access law might pass judicial review. Professor Jerome A. Barron in his book, *Freedom of the Press For Whom?* believes that such a statute would not abridge a newspaper's First Amendment rights by allowing access.[1] A new right would be created without limiting existing rights. A newspaper could continue to print what it pleased, attack whomever or whatever it chose as has been its right in the past. Any person or group that was attacked, however, would have the right to buy space and offer a defense.

Assuming that such a statute were held constitutional, which is far from certain, the social and political implications would be profound, especially so in one-newspaper markets.

Many of the more progressive newspapers have recognized the need for more diversity of opinion in its editorial page (the Op-Ed page of the *New York Times* being an example). This represents a policy change; it is access, but it is not a *right* of access. This discretionary access might be characterized as significant tokenism — to be distinguished from the authentic tokenism practiced by the broadcast media.

In allowing access, there is one important difference btween the print and the electronic media: the former can always add a page or two to its publication; the latter are limited by the number of broadcast hours. This is no way excuses broadcasters from doing as little as they do, limiting access in favor of more profitable pursuits.

What we are experiencing in broadcasting is closely related to the early entry of newspaper owners into radio. They brought to this new medium the mind-set and attitudes, the prerogatives and absolutism associated with publishing; the idea of power sharing, licensing, and constraints was foreign to their experience: many considered regulation odious and burdensome, a corruption of the natural order of things. It was difficult for some to accept their new role as public trustee or to take it very seriously. It was an impediment, something to be ignored or circumvented; it would be accorded lipservice and little more. Regardless of what the Communications Act had to say about public trusteeship or the public interest, they intended to operate as they pleased; that is to say, in their own self-interest with but an occasional recognition of their public trust. This attitude pervades broadcasting to this day, and no institution in or out of government is able or willing to do anything to change it. The take-over is complete. We are dealing with established power.

A seasoned politician knows where the levers of power are and has most likely made some accommodation; the aspiring politician must do the same or accept an uphill struggle. Political candidates inherit a world not of their own creation; many would prefer it otherwise, while others feel no discomfort serving narrow interests. Faced with reality, a politician tends to become pragmatic; idealism is abandoned in favor of realism. The "logic of the facts" asserts itself; the rationale becomes a political verity. To be elected one needs votes; to gets votes, one needs media exposure, especially favorable exposure; to obtain this support, one must adapt one's politics to meet the minimal demands of the mass media. Some politicians are so well established that they can break all the rules and survive; they are, however, the exceptions that prove the rule, the rule being that in swing districts where elections are won or lost by a few percentage points, media support is often decisive.

To the casual observer, television — in contrast to the more partisan press — tends to be a bland, neutral medium. The long knives are seldom, if ever, sharpened on camera; dispatching an unfriendly or unacceptable politician is accomplished with style and sophistication; it has become a minor art form, particularly suited to the medium. There are several approaches, one being simply to ignore a candidate, creating a medium nonperson, the idea being that if one is not on television, one does not exist — politically speaking. Newspapers have used this method of dealing with their enemies, often with telling effect. Voters begin to wonder about the effectiveness of an elected official; they seldom consider a conspiracy of silence. By implementing a policy of neglect, the mass media can manipulate and distort the political process.

The activist mode also involves conscious decision making, one that employs selective coverage, the net result always being a political minus for the victim. What may appear to the casual observer as the essence of objective electronic journalism — words straight out of the politician's mouth — can be in the hands of an unfriendly public trustee, a potent weapon. If, for example, a politician makes an hourlong speech and only thirty seconds of it will be used on the nightly news, the question then becomes, Which thirty seconds? This is and ought to be a news judgment; but in partisan hands, it may have the devastating impact of an editorial comment — and a secret one at that.

Under the best of circumstances, television makes great demands on any politician; the camera is a merciless observer, catching subtle nuances of meaning and motivation, frequently revealing more than the subject intended. Only a gifted few master the medium — most handle it poorly; some not at all. If with all the inherent demands imposed by the medium, a politician must also protect himself from editorial emasculation, Is it any wonder how few provoke the wrath of the gods?

The enormous power of television is not abused by all public trustees; many prefer to stay away from the political wars and concentrate on making money. Nevertheless, there is a club in the closet ready to be taken out and used against any unfriendly congressman or senator who fails to vote "right" on legislation affecting television's vital interests. This is, of course, an implicit threat, one that has not escaped the attention of those interested in political survival. Invoking the displeasure of the media can prove risky, and if there is one thing politicians avoid it is taking risks, especially unnecessary risks. To get along with the media, one must go along.

4

Delivering the Message

It may only be coincidence that the era of the strong or imperial presidency came into being at the same time that radio emerged as a mass-communications medium. To be sure, there were other factors: the Great Depression, World War II, and with it the age of nuclear power — developments that weakened the sharing of power between the Congress and the President. To have 535 fingers on the nuclear trigger was not a viable option; technology and events mandated a delegation of power — including as a practical matter the power to wage nuclear war — to one person, the President. Congress, though a coequal branch of government, thus became relegated to a distinctly secondary role.

The exalted role of the President was nourished and to some extent created by electronic mass communications; popular perceptions could be influenced by the President who could directly address the people himself, bypassing the print media. It was the era of selling a presidential image through a medium he could control. Skillfully employed, radio became an instrument of power. The disembodied voice of the President was reassuring in times of trouble; people believed in him because they needed and wanted to believe. He became for many the symbol of the nation itself. When he spoke it became almost unpatriotic to question his motives or judgment. Once the flag had been planted in some foreign land, we followed.

It was not always that way. In the nineteenth and early-twentieth centuries, Presidents had a more human dimension, a less-inflated public image. Many were scorned and ridiculed in their own time; even the great ones — George Washington excepted — had to die before they were properly appreciated. In those rough-and-tumble days, Presidents were treated rather harshly — like any other politician. Congress was anything but deferential, jealously protecting and asserting its constitutional prerogatives as a coequal branch of government.

The influence of the electronic media in the erosion of congressional power is difficult to measure — that it was, however, part of the process is clear. Though still powerful, the print media became a technological casualty; no longer did the press have a monopoly in mass communications; no longer did it have sole control and power. At a time of his own choosing, a President could take to the airways and deliver his message to millions of Americans, thus circumventing the press. The press lords, accustomed to playing the kingmaker role in politics, had to share power with this electronic interloper. In consequence, the press had to adopt somewhat higher ethical standards — particularly in the area of straight news reporting — or suffer a loss of credibility. Radio could give the lie to inaccurate or slanted journalism; politicians now had an alternative medium.

With the advent of radio, the political process was fundamentally changed in other ways. The leather-lunged orator on the hustings had become obsolete. With electronic assistance, an ordinary human voice was deemed adequate, particularly so if the speaker could project the desired image. The new breed, like their progeny a generation later, had to adapt their style to the requirements of the medium. As a result, some political fortunes rose while others faltered; the tyranny as well as the opportunity of the medium became a political fact of life.

Unlike Congress, the President had access to radio. This was a matter of policy, not a matter or right. The networks had no legal obligation to make time available to anyone, the President included. As a practical matter it was deemed prudent to give a President the use of the public airways when he wanted to address the nation as President, as distinguished from political candidate. To do otherwise was to risk presidential disfavor. If sufficiently aroused, he just might appoint unfriendly people to the Federal Communications Commission or propose undesirable legislation. The broadcasters might be able to block this in Congress, but being sensible businessmen the better part of wisdom suggested accommodation rather than confrontation.

Congress was treated differently. Access was entirely a matter of discretion to be exercised by the networks and their affiliates. Although a coequal branch of government with a need to communicate with the electorate that was no less important than that of the President, Congress as an institution and its members as individuals were not granted access. Broadcasters would decide when, whether, and how much time was to be made available. Clearly Congress had the power to change this. What it lacked was the political courage.

The net result was that the President could go directly to the people and try to sell his program; congressmen and senators were obliged to pass through the mass media filter.

Thus, through broadcasting fiat, distortions in the political process came to pass, the President becoming stronger and Congress weaker. Understandably, not everyone in Congress was overjoyed by the new realities that accompanied this technological change. Despite the grumbling, Congress waited for some forty years and the advent of television before agreeing to study the matter.

Finally, in December 1972, Congressman Jack Brooks, chairman of the Joint Committee on Congressional Operations, requested the Congressional Research Service to investigate how Congress might use the mass media more effectively. The Congressional Research Service commissioned John G. Stewart to make a study and write a report. In the preface to this report, congressmen Jack Brooks and Lee Metcalf stated their case.

Congress, in exercising its responsibilities as a co-equal [sic] and independent branch of the Federal government, has a substantial stake in being able to communicate with the American people effectively.

This is particularly the case in light of the massive and highly sophisticated use of mass communications by the President and the Executive branch. Whether measured in numbers of people or by annual budgets (approved by Congress), the capacity of the Executive to communicate its views is staggering. And this capacity grows relentlessly with each new administration, contributing at the same time to the growing imbalance between executive and legislative power.

Moreoever, the President possesses a unique potential for dominating the communications media and, ultimately, public discussion of critical policy issues. Most susceptible to such domination is the most powerful of the media — network television — which projects the presidential image and voice into the living rooms of most American families. By virtue of his office, the President is routinely granted prime television time by a simple request to the three commercial networks. But the Congress, even if it is controlled by the party which does not control the White House is given no such routine access.[1]

The report documents the use and abuse of presidential access to television while Congress, a coequal branch of government, is reduced to the role of media supplicant, passively awaiting the attention of network news departments. This institutional disparity in access of the electronic mass media is bad public policy, the report argues.

It is not only the President who employs the mass media as an instrument of power. In the executive branch of government there are hundreds of public-relations employees whose job it is to inform — or propagandize — on behalf of an agency or program. Ironically Congress, which complains about its communications disparity vis-a-vis the executive branch, routinely approves funds for those positions. Consequently, when a federal program or agency needs to be cut or abolished, Congress must do battle in the arena of public opinion with the very public-relations experts that it itself created.

In 1971, the Office of Management and Budget reported that the federal government employed 6,144 people working full-time in public relations. This number represented only the identifiable employees.[2] "Today," the report concluded, "the smallest and most inconsequential bureau in the Executive Branch is quite likely to have at its disposal a press and public-relations capacity that exceeds the institutional resources available to either the Senate or House."[3]

To remedy this media disparity, the House saw fit to allow television coverage on a restricted basis. In February 1979, an inhouse closed-circuit television system, employing remote-controlled cameras, was activated. There were cameras covering the Speaker's rostrum and the tables of the minority and majority leaders. Cameras were not permitted to pan the aisles, or show the vacant seats. There were no minicams roaming around, no reporters asking difficult questions.

Even this timid use of television encountered congressional opposition, some predicting that a circus atmosphere would result with congressmen playing to the cameras and their constituents back home.

This inhouse use of television with its many constraints made for rather dull viewing. None of the networks, including PBS, now broadcasts the proceedings on a regular basis, although C-SPAN, a cable service, does.

5

Advertising: Where The Money Is

While television is supposed to be "free," it has in fact become the creature, the servant and indeed the prostitute of merchandising. — Walter Lippmann, 1959

Commercial broadcasters are fond of perpetuating the myth that they are providing a free service to the American people. What they actually provide is a funding mechanism in concert with advertisers who in turn pass the costs along to consumers in the form of higher prices. The system is, in effect, a private tax which since it is hidden seems painless. We appear to be getting something for nothing, but such is not the case. There is no free lunch. Everyone has to pay.

Some of the larger advertisers argue that their huge advertising budgets are absorbed in the economies of scale; i.e., advertising generates sales, sales generate increased production with resulting lower unit costs. It is a plausible theory but one that cannot stand close scrutiny.

A study, by Harry J. Skornia, on four nationally advertised products — Johnson's Wax, Revlon, Hazel Bishop, and Alberto Culver — reveals that as the product gains popularity resulting from advertising, the price went up, not down.[1] Having established through advertising the belief that their products had a qualitative advantage over the competition, the manufacturers were able to charge higher prices. Thus the consumer unknowingly pays a double tax. No doubt, there were cost savings resulting from increased production; these savings, however, were passed along to stockholders, not to consumers.

For many advertisers, television is absolutely indispensable to sales and earnings, a cornerstone to their economic well-being. There are many success stories of insignificant companies rising to prominence by having understood and exploited the medium. Oddly enough, it is not a matter of having a better mousetrap and spreading the word; more often the mousetraps are either identical or only marginally different. The task of advertisers is to convince an uncritical and unsuspecting audience that there are qualitative differences. When the public buys the product, the cost of this deception is included in the price; if many people are persuaded, the cost of the product may rise even higher. There are many products that fall into the category of having little or no difference in quality, despite advertising claims to the contrary: aspirin, soap, toothpaste, vitamins, deodorants. What these and other products have in common is that they advertise extensively on television. *The less difference in a product, the greater the need to claim otherwise.* Television did not create deceptive advertising, but it has sustained and legitimized it. And therein lies a moral dilemma and a fundamental question of public policy: Should our public property be used for disseminating deception, half-truths, and not infrequently, outright lies? Not only is the public victimized, it is further required to foot the bill. If the cost is calculated in economic terms — an appropriate criterion — billions of dollars are involved every year. Cost though important ought not to be *the* most important factor in a society that purports to value truth above all things — including product advertising. Do we not have the right to insist on the highest standards of truth? Is it sound policy to allow advertisers and public trustees the use of our airways to sell products that are useless or harmless? Despite hundreds of millions of dollars expended annually to promote patent medicines, the fact remains that scores of them are entirely useless, overpriced, and in some cases, dangerous. Thanks in large part to advertising on the electronic media, we have given birth to a pervasive drug culture where a pill can be found to solve almost any problem. Not everyone has been victimized by this dangerous nonsense, but there is ample evidence to link permissive advertising to some very serious health problems.

The FCC, the Federal Drug Administration, and the Federal Trade Commission have shown little enthusiasm for taking on the powerful advertiser-media combination. When aroused to action by a demonstrable outrage, the result is generally prolonged litigation that does little or nothing to solve the basic problem. With no effective opposition, advertisers continue business as usual.

The people who produce advertising — the advertising agencies — come in for a great deal of criticism, some of it justified. But it should be remembered that they are the hired hands rather than independent decision makers. The real power and ultimate responsibility rests with those who pay the bills — the corporate clients. Where advertising on the public airways is concerned, it is very much a case of buyer beware.

Advertising per se is not the issue. In a complex system of production, distribution, and consumption, advertising serves a necessary economic and social function. Accurate information fairly presented is one matter; conscious deception is another. One might think that the owners of the spectrum would have a right to expect the highest ethical standards from those privileged to serve as public trustees. Such expectations, however, seem totally unrealistic and naive given our experience of the last five or six decades. We have all been conditioned to expect — even to accept — deception on the public airways. It is as American as apple pie.

Under the law, licensees can refuse to broadcast *any* commercial not to their liking; their discretion is absolute. Whatever ills exist in commercial advertising can be remedied by broadcasters. They have the power to change things, to upgrade standards, but it is not in their interest to do so. It would be bad for business if advertisers were obliged to deal factually and truthfully with the general public. In such an antiseptic environment, many advertisers might choose another medium or not advertise at all.

Selling to an adult audience by employing all the tricks of the ad man's game is one thing; using these same tactics on an audience of children is another. The former can exercise some reason and judgment, discounting some of the more blatant deceptions. Children are infinitely more vulnerable by virtue of their lack of maturity, experience and judgment. They are an easy mark, part of a huge market delivered to the hucksters of commercial television every Saturday morning. Though short of money, children can be very persuasive in getting their parents to correct this deficiency, which is, of course, the whole idea. The target is the parent; the child is merely the messenger.

This assault on the innocent mind poses a moral question within a commercial context: Should any child be manipulated and exploited for profit? The question is relevant despite decades of precedent. Is it really in the public interest to allow advertisers unrestricted access — often without any parental involvement — to sell their merchandise to children free of meaningful constraints?

The economic implications of changing the rules on television advertising are enormous; the kiddie market is huge. But measured against the ethical considerations, commercial imperatives ought to be of secondary importance if, as we claim people are more important than profits.

If television is a teaching machine, what do children learn from commercials? That truth itself is the first victim of the medium, that it is held in low esteem, and that, inasmuch as adults on television lie for money, children like the rest of us should be armed with a certain amount of cynicism. Thus television teaches the child to deal with the real world of commerce, competition, and survival — another public service.

Broadcasters argue that they have no choice other than selling products to children; somebody has to pay for the cartoons. This reasoning is a good illustration of using the system to justify the system — that what exists is justified by the very fact of its existence. This is a simplistic and disingenous non sequitur. The decision to exploit children on television is a conscious policy of advertisers and broadcasters; nothing in the Communications Act or the FCC rules mandates it. Broadcasters could, if they so chose, underwrite children's programs out of their huge profits; they could raise their rates slightly on other advertising to support such programming. They could, but they won't, for the very good reason that they don't have to. And why should they? There is no economic incentive for change; indeed, all incentives favor the status quo.

If Congress wanted to, it could change the rules of the game, but there is no hope that it will; the broadcasting lobby would be certain to kill it. Getting a bill through Congress that adversely affects the industry's profits is next to impossible; it has not happened yet, nor is it ever likely to happen. Expecting Congress to act against these powerful vested interests is unrealistic. It is equally clear that we cannot expect fundamental change from within the industry. Like most institutions, television is not inclined to question its prerogatives. We can hardly expect one businessman — our public trustee — to challenge the interests of another businessman — the corporate advertiser — when the latter wants to make the former rich. Their interests are in harmony. They live off each other. The result of this congenial relationship is that the public airways have become an advertiser's showcase.

6

Entertainment: Giving The Public What It Wants

We have triumphantly invented, perfected, and distributed to the humblest cottage throughout the land one of the greatest technical marvels in history, television, and have used it for what? To bring Coney Island into every home. It is as though movable type had been devoted exclusively since Gutenberg's time to the publication of comic books. — Robert M. Hutchins, 1961

The standard industry response to criticism on the quality and diversity of commercial television programming is that it has given the public what it wants and the proof lies in the ratings — the holy writ of the broadcasting industry. The explanation appears plausible, the numbers impressive. Clearly a lot of people have watched and are watching commercial television. On closer examination, however, what appears to be the quintessential manifestation of the popular will is in reality yet another business decision.

One must first define which public we are considering. Are we talking about old people living on modest fixed incomes? Or are we referring to the middle-aged and younger segment of the public? Does this include preschool children? There are many publics to be served, all with different needs.

While broadcasters claim to "give the public what it wants," it might be more accurate to say that they "give advertisers the public they want." In television as in life, some people are more equal than others. This is especially so where advertisers are concerned; they are interested in a particular public — those people with sufficient disposable income to buy their products. For advertisers, selling big-ticket items — automobiles, appliances, etc. — the size of the audience is less important than its demographic profile. Elderly people living on Social Security do not buy new cars and refrigerators with the same frequency as the younger, more affluent segments of the public do. In consequence, programming decisions cater to a particular audience, the audience advertisers want to reach.

When pressed, broadcasters will concede the obvious, that indeed the requirements of advertisers are taken into account in program decision making. And why not? They pay the bills. Television, they say, cannot please everyone; it cannot be all things to all people. And if we broadcast trash a lot of people are watching and we have the numbers to prove it.

The fact that people watch what the networks offer does not necessarily support the propostion that they especially like what they see on their screens. The numbers may only reflect the popularity of the medium, that people will watch something — anything — rather than turn the set off. It is, perhaps, less a question of enthusiasm than of boredom, a matter of passing time. Given the choice between something and nothing — a blank screen — many choose the former.

With three commercial networks, some critics ask, why is there not more diversified programming? Why are offerings so remarkably similar: cop shows, medical shows, situation comedies? Why is there no attempt at diversification with one network striving for excellence, a second taking the middle of the road, and a third catering to less demanding tastes? The answer, of course, is money. Every network has taken dead aim at the same audience in a determined battle to win the numbers game, thus increasing the value of its advertising time.

To suggest that broadcasters ought to be cognizant of their obligations as public trustees and make an effort at quality and diversity is a plea that goes unanswered. The public-service rationale for broadcasting has yielded to the imperative of profit maximization. Nothing supersedes the effort to win the ratings war. The dynamics of network television are such that no force from within has the power fundamentally to restructure the system. The system is self-contained, securely harnessed to the money machine, programmed to make all it can. The American people who never asked to play a numbers game or become involved in a ratings war are victims of the system that was supposed to serve them.

Complaints about the quality of network programming elicit the following response: We do not create public taste; we respond to those tastes. If the public wants higher quality programs, we will be glad to supply them. On those occasions when we have made an effort to upgrade quality, the people do not watch.

Implicit in these seemingly reasonable arguments is the condition that quality must attract huge numbers of people or perish, that those who want quality are asking for too much if networks and their affiliates end up making less money for their efforts. Asking the networks and their affiliates to strike some reasonable balance is similarly unacceptable.

The question of responding to rather than elevating the public's taste merits inquiry. Nobody would support a forced feeding of "good" programming; one man's cultural feast is another man's garbage. We neither need nor want an elitist diet. What we have a right to expect is diversity, more choices for more people — and that out of this diversity, some people's taste will have an opportunity to evolve. At the very least, that option should be provided. The lowest-common-denominator approach to programming is good business for the broadcasters, but it is not good public policy. There are millions of people — though perhaps a minority — who want quality and diversity, a range of choices that those who control the medium refuse to provide. And why should they when they can make more money selling trash? There is no incentive for quality.

Because of its pervasive influence, its presence in nearly every American home, television has had more than a few critics. The movers and shakers of the medium have come to accept criticism as the price of doing business, a burden to be endured. The public relations of broadcasting dictates a reasoned response to criticism, not an abrupt, arrogant dismissal. Public trustees must *appear* to be receptive to legitimate complaints.

Nothing better illustrates the sophisticated approach than the periodic public protests about violence on television. It is the medium's longest-running soap opera. Industry critics, politicians, regulators, and broadcasters interact in a morality play, the outcome of which is entirely predictable. Violence on television is as old as the medium itself, a problem that has always plagued the industry. It is not that broadcasters as a group have any special love for violence. It is just a matter of business, something that has to be done to move the merchandise. Programs with a high level of violence do well in the ratings, and broadcasters therefore feel obliged to exploit it. Besides, they argue, violence is a part of life, and television should reflect reality, not sanitize or distort it. From the time of the classical Greek tragedians to William Shakespeare and on into the modern era, violence has been a part of drama. Why should television be any different?

The issue is not reflecting life as it is or following dramatic precedent; rather it is the gratuitous and exploitive use of violence for the purpose of raising the ratings, which produce larger profits. Greed, not art, is the motive. Broadcasters, driven by the competitive requirements of the ratings war, feel compelled to mine the darker side of human nature. It is a dirty business, but it makes money.

From time to time, there is a public reaction to this shameless exploitation and demands for change. When protests grow loud enough to reach the congressional ear, an investigation is initiated. The scene is by now familiar: an articulate network spokesman seated at the witness table surrounded by aides and lawyers. The chairman bangs the gavel, throats are cleared, and the show begins. The network representative reads a prepared statement that is persuasive, reasonable, reassuring. Yes, his network is fully cognizant of the problem and is doing its very best to prevent excesses. No, they do not program violence for the purpose of getting higher ratings. Yes, we will keep an eye on the matter and will try to do better. To the suggestion that Congress might be prepared to impose a solution, the network representative calls forth the spectre of censorship and its many evils. The interrogator beats a hasty retreat. Nobody wants censorship; on this there is full agreement. Words like "responsibility" and "self-regulation" are heard. And so it goes. The networks have been publically chastised and warned to mend their violent ways. Congress has acted; it has done something about TV violence.

With the media event concluded, the industry can take satisfaction in once again protecting its prerogatives to broadcast whatever it chooses. There are, of course, self-imposed limits that broadcasters adhere to. Like all sensible businessmen, they have to have a good understanding of the customer and the product to be successful. For broadcasters, this is their area of expertise; knowing what the public wants or is prepared to accept remains an essential part of the business. When controversy and protest surface to the level of political scrutiny, it means somebody has badly miscalculated the public mood, an inexcusable error for professional communicators. Only when there is a lot of money at stake as in programming violence will broadcasters take risks. Ordinarily they tend to prefer bland, inoffensive offerings designed not to arouse the natives.

Broadcasters, not unlike bankers and multinational corporation executives, like to underplay their real power — a stance that comes not from inherent humility. It is a calculated business judgment. Arrogance not only is bad form — indicative of an absence of sophistication — it also is bad for business. A "public be damned" attitude is simply out of the question. For broadcasters, nourishing the mythology of accountability is vitally important. To reveal their real power — let alone to flaunt it — would be a serious error. When attacks on the industry center on peripheral issues such as sex and violence, broadcasters count their abundant blessings, grateful that there is no frontal assault, and confident that everything else is manageable.

The most persistent criticism of television is the effect that violence has on children. The industry argues that there is no clear correlation between what its critics contend is violent programming and antisocial behavior. The evidence is, they argue, at best ambiguous and often contradictory.

A minority of authorities tend to agree. Some even suggest that exposure to violent programs is cathartic, allowing the child to work out his anger, frustration, and latent hostility.

The majority view — sustained by more than twenty-five hundred studies, a cottage industry for sociologists and psychologists — finds a cause-and-effect relationship, that exposure to violent programming does indeed encourage antisocial violent behavior.

When experts disagree, laymen are forced to make their own judgments, to exercise a large measure of common sense. Can anyone seriously argue that a child daily exposed to six hours of television will not be affected in some way? Are we to assume that these impressionable minds are impervious to the pervasive violence, the hundreds — indeed thousands — of shootings, muggings, rapes, and murders that they witness? Broadcasters say no harm is done.

Some parents make an effort to control their children's access to television. It is no easy task. Nothing less than monitoring every program is entirely satisfactory, something that most parents are unwilling to do. Many fail to see the risk, the assumption being that if television was harmful to children somebody in authority would put a stop to it. For some parents, television is the best invention since the pacifier. It keeps the kids quiet and out of the way for a few hours; it is also educational, molding values, communicating ideas, preparing the young for the "real" world they will soon inherit. Despite all the studies nobody truly understands the consequencs of this electronic conditioning. Given the profound complexity of human behavior, it is often difficult to make precise correlations between exposure to violence and violent acts. How we have become as we are and act as we do remains mysterious, fertile ground for philosophers, theologians, and social scientists to till.

The commercial hucksters from Madison Avenue and their corporate clients have no doubts concerning their ability to influence children's behavior, a conviction backed up by hundreds of millions of dollars annually in advertising dedicated to persuading children to *do* something, i.e., buy the merchandise. And it works; children do respond to advertising messages, and advertisers have the sales records as proof. That a professional communicator can persuade a child to do something he wants him to do is in no way surprising. It is, of course, a mismatch. At the same time, and by some tortured logic, we are asked to believe that the programming that fits between the commercials does *not* affect children's behavior, that there is no cause-and-effect relationship associated with violent programming and violent behavior.

Advertisers and broadcasters are trying to have it both ways by contending that commercials do persuade, do alter behavior, but that programming doesn't, this despite the fact that children might be more susceptible to programming messages — intended or not — than commercials. This dichotomy of convenience is the operational assumption of violent programming. Under the present system and prevailing power relationships, it seems very unlikely that anything will fundamentally change.

7

Working Within the System:
Edward R. Murrow and Fred W. Friendly

Look, Fred, I have regard for what Murrow and you have accomplished, but in this adversary system you and I are always going to be at each other's throats. They say to me (he meant the system, not any specific individual) 'Take your soiled little hands, get the ratings, and make as much money as you can'; they say to you, 'Take your lily-white hands, do your best, go the high road and bring us prestige!' " — James Aubrey, president of the CBS Television Network, talking with Fred W. Friendly, executive producer of "CBS Reports"

The careers of Edward R. Murrow and Fred W. Friendly, employees of CBS, are instructive case studies on two talented professionals trying to work within the commercial broadcasting system. Their ultimate disenchantment had less to do with personalities within the CBS establishment than with the imperatives of the industry, one dedicated to profit maximization, a self-sustaining dynamic with a life of its own. The moneymaking machine must be kept running at top performance; the people who own the machine — the stockholders — demand results. They are not interested in vague abstractions such as the "public interest"; that is the stuff of public relations. Nobody was immune, not even CBS founder and chairman of the board, William S. Paley. The business of broadcasting was to make money, and everyone was expected to do his duty. Those found wanting in this regard were quickly dispatched.

Murrow as journalist and Friendly as producer were not rebels trying to cut down the corporate tree. Both were employees of long-standing, accustomed to the constraints inherent in large organizations. Murrow had begun working for CBS in 1935 and remained in its employ until 1961. Friendly started in 1950 and stayed for sixteen years. By network standards where people come and go quickly, their longevity was exceptional. Had they been chronic troublemakers demanding their own way all the time, they would never have lasted so long.

Both Murrow and Friendly had high standards of journalistic integrity. Murrow was the foremost electronic journalist of his time, which gave him a certain latitude to do and say things that others could not. Still, he was obliged to walk a thin line, remaining faithful both to CBS and himself. It was not always an easy task.

When television came of age in the early 1950s it needed intellectual prestige to go with its enormous popularity. Milton Berle might move the merchandise and make a lot of money, but he could not sell ideas. For this endeavor others were engaged. CBS turned to Edward R. Murrow, a familiar name and voice, if then, a still unfamiliar face. CBS had not only a fine journalist in its employ but a world celebrity, a man who had known and been admired by Roosevelt and Churchill during World War II. Murrow was a corporate asset, and CBS wanted to use him to add stature to the infant medium of television.

Murrow was willing to give it a try. The first "See It Now" program was telecast in November 1951, with Fred Friendly as producer. With characteristic honesty, Murrow looked into the camera and conceded his unease: "This is an old team trying to learn a new trade."[1] It was a struggle, and it showed. Murrow looked stiff and uncomfortable; he could not relax; the searching, unrelenting, remorseless camera would not permit it. The medium made demands upon him quite different from those of radio where his superb speaking voice and considerable eloquence sustained him. Television is a visual medium that accentuates how a person looks at the expense of what he has to say. Murrow had a lot to say, and he said it well, often movingly, but he seldom looked at ease saying it. That he survived and enlarged his journalistic reputation despite — not because of — the medium is a triumph of ideas over images. The superficiality of television yielded to substance; Murrow would have it no other way. The medium would not determine the message.

A serious man, Murrow took his work seriously. He refused to pander or talk down to his audience. His assumption was that his viewers were prepared to think about what he had to say, that they were interested in the world and its problems, rather than escape into the vast wasteland. Encouraging people to think was something new for television — and a little dangerous. People who thought might come to regard television as an instrument of enlightenment, and might eventually demand that the medium meet a higher standard of responsibility. An informed public represented a threat to the status quo in which broadcasters and advertisers had a vested interest. For those who ran it, television was a moneymaking enterprise; for those who watched it, a diversion, an escape from the concerns of the real world. Thoughtful people could go to the public library.

For CBS, Murrow and Friendly were something of a mixed blessing. They were needed for prestige, to protect the corporate flank from attack; an insurance policy, and as such, part of the corporate mix. To make credible its commitment to news and public-affairs programming, CBS was obliged to do more than issue a press release; it had to have some factual basis; to flesh out the desired image there necessarily had to be substance as well.

In creating a strong news and public-affairs department, CBS assumed certain risks. Journalists — at least, the good ones — assert a measure of independence in their work; an employer cannot dictate content, only general-policy guidelines. Overtly to intervene in the reporting or editorial process would compromise journalistic integrity which could lead in turn to exposure and resignations. If a journalist was to be controlled, it had to be done discretely, not by fiat.

For a journalist like Edward R. Murrow, direct control was out of the question; he would not stand for it. A sophisticated organization like CBS would never resort to such a dangerous option; there were other less risky means of exercising control. First among these options is the decision to fund a program. A second involves time: How much and when? Will a documentary be given a half hour or an hour? Will it be aired in prime time or consigned to time when few people watch? These are corporate decisions beyond the control of journalists, even famous ones.

Once funded and given a time slot, Murrow and Friendly exercised control over content — a grant of authority that today is very much the exception. Group journalism with significant management oversight is the rule.

Electronic journalists and their news departments are nominally independent, free to report the news exercising professional judgment. This is certainly the case most of the time on most of the stories, but it would be naive to assume that reporters do not bend to the prevailing wind, accepting inhibitions and self-censorship. They are, after all, like Murrow and Friendly, corporate employees who serve at the pleasure of management. Doing the wrong story in the wrong way, challenging powerful interests groups, can be hazardous to one's career. It was true in Murrow's day, and it is still true.

Murrow and Friendly came to television when it was still in its infancy, before all the rules were carved in concrete. They fought for their independence and used it, the result being the very best, memorable, and bravest television broadcasts ever seen.

A deliberate man, Murrow did not come to the new medium breathing fire and looking for devils. Nor was he looking for the big story, the flashy expose. He was an intellectual, interested in ideas and the broader implications of those ideas. Radio was his medium, an instrument more accommodating to the world of ideas, less distracting to thought, than television, a medium overwhelmed by pictures. Yet television was where the action was; radio was in decline. Reluctantly Murrow signed on. Television was a challenge, one he could not ignore and remain a journalistic force. To reach the people he wanted to reach, to explore the issues he considered important, he would have to embrace the new medium.

Murrow was a courageous journalist, one not inclined to avoid a controversial story just because it might cause him or CBS problems. Exploring controversial issues was a professional responsibility. Yet he could be cautious. It took him nearly two years of "See It Now" programs before he confronted the most explosive issue of that period — McCarthyism. Senator Joe McCarthy had created a climate of fear and suspicion that pervaded the whole society. It was an issue that demanded examination, but few journalists wanted to face the senator and his legion of followers. McCarthyism was too controversial, too explosive. A newspaper column or broadcast might end a career. At the very least, its originator would become the target of abuse; his reputation would suffer; his judgment, even his loyalty, would be questioned. It was a moment of truth for many a journalist, and most of them let the moment pass. Not so Murrow. He had been waiting for the right moment, the right story, to address the issue, and finally he found it in the person of Lieutenant Milo Radulovich. Lieutenant Radulovich, age twenty-six, a meteorologist in the United States Air Force Reserve, was losing his commission. He had been judged a security risk by reason of his having maintained a close and continuing relationship with his father and sister who allegedly read subversive newspapers and whose political activities were deemed to be questionable. The loyalty of Lieutenant Radulovich himself was never in question; he had been adjudged a security risk only by virtue of his relationship with his father and sister. It was an interesting question with the most profound implications; it deserved a public airing.

On October 20, 1953, Murrow looked into the camera and made a fateful declaration: "We propose to examine ... the case of Lieutenant Radulovich ..."[2] It marked the beginning of an era; electronic journalism had come of age; it was dealing with issues — controversial issues. Murrow was acutely aware of what he was getting into. "I don't know whether we'll get away with this or not," he had remarked to Fred Friendly moments before the broadcast. As things turned out, he did get away with it. The response from the public was overwhelmingly favorable; people were interested in justice, in due process and elementary fairness. Television could do more than entertain and sell merchandise: It could make people think; it could inspire; it could change our lives.

The management at CBS maintained a discreet silence. No one said a word to either Murrow or Friendly before or after the telecast, but Friendly detected some "uneasiness" in the CBS high command. CBS had been presented with a fait accompli; it was too late for protest. Had they admonished Murrow and prohibited him from engaging in this kind of controversy in the future, the whole concept of journalistic independence — an avowed network policy — could not have survived particularly if Murrow and Friendly made an issue out of it by going public. This was not the kind of risk that CBS was willing to run; it might well have been a public-relations disaster. Although the business of television was and is to make money, the concept of rendering a public service — particularly in the area of news and public affairs — had to be maintained. Though burdensome and potentially harmful to CBS's corporate interests, Murrow and Friendly had to be tolerated — at least, for a time.

On March 9, 1954, Murrow again tempted fate, this time by openly taking on Senator McCarthy. It was a struggle between two heavyweights, each operating from a different power base. Murrow was aware that it would not be a civilized exchange of views, rather a no-quarter struggle — at least on the senator's part — leaving a lot of blood on the floor. McCarthy was an awesome, intimidating, powerful foe who generated fear in the bravest of men. Very few stood up to him, and those who did often lived to regret it. A talented demagogue with an instinct for the jugular, McCarthy did not play favorites or accept any inhibitions. Everyone was fair game: a low-level bureaucrat, an army general, a Secretary of State, even a President of the United States. They were all the same to Joe McCarthy.

The decision to do a "See It Now" program on Senator McCarthy was not made lightly; both Murrow and Friendly understood they were risking their careers and reputations on the outcome. McCarthy would strike back with all his fury; of that, there was no doubt. If they waited any longer, with the senator's growing power, he might be impossible to stop. A climate of fear had taken hold of the government, the press, the radio and television networks, the commentators, writers and actors in movie studios and elsewhere. There was a Communist lurking in every closet; people were ending up on blacklists, their reputations and careers ruined without benefit of evidence or a trial. This mass hysteria had but one antidote — the truth — and someone with the courage to speak out. Murrow and Friendly decided to give it a try.

The CBS management knew about the McCarthy program but was in no way involved in the decision to do it. Murrow exercised his own journalistic discretion even on a story that might well shake the network to its foundation. All CBS could do was to hold its breath and hope that everything turned out all right. Intervention — the only other option — involved unacceptable risks.

The telecast was followed by an avalanche of telephone calls, most of them favorable to Murrow. Thousands of letters and telegrams were received by the network supporting Murrow by a ratio of about 10 to 1. It was an encouraging development.

CBS adopted a policy of silence except for the issuance of a statement defending Murrow's integrity and patriotism; that it felt obliged to do so in the exercise of free speech was noteworthy. There was no mention of the broadcast. CBS obviously wanted to put some distance between itself and Murrow as it awaited events. If Murrow was to go down, CBS was not inclined to share his fate — at least, not voluntarily.

The private view of the network establishment was summarized by Murrow after attending a CBS board of directors meeting: 'Good show, sorry you did it.'''[3] As a work of electronic journalism it was a triumph; as an adjunct to the business of broadcasting, it was a potential disaster. How many such victories could CBS have and still survive?

Frank Stanton, president of CBS, was worried. He called Fred Friendly into his office to show the results of a public-opinion poll he had commissioned Elmo Roper to do on the controversy created by the McCarthy broadcast.* Fifty-nine percent of the population had either seen or heard about the McCarthy program and of these 33 percent believed that the senator had proved Murrow a pro-Communist or had raised doubts about him. This was in Stanton's view alarming news; Murrow had made himself — and CBS — controversial. It was an unforgivable sin; it was bad for business. That twice as many people believed McCarthy had failed to make his case against Murrow was beside the point: a substantial minority thought he had. All this could have been avoided if Murrow had had the good sense to avoid sensitive issues. From a corporate viewpoint, it did not matter that the debate centered on the most fundamental principles that sustain our democracy: due process and free speech. As an individual, Stanton probably agreed with Murrow; as president of CBS, he could not. His job was to make money — all the money he could. Murrow was making that task more difficult.

CBS had a problem. It was riding on the back of a tiger; getting off or staying on might prove equally dangerous. It decided to finesse the issue — to do a little of both. "See It Now" was cut back along with the total autonomy Murrow and Friendly had previously enjoyed. The broadcast would be replaced by eight or ten irregularly scheduled "See It Now" programs. The very considerable power that Murrow and Friendly had enjoyed to produce what they pleased without consultation with management was now lost. Though still enjoying considerable discretion in the selection and production of programming, Murrow and Friendly had become part of the CBS news team. It marked the end of an era. Never again would anyone have the kind of freedom that Murrow and Friendly had had.

Ultimately "See It Now," the most honored public-affairs program on television was canceled. CBS, now firmly established, no longer needed the prestige or the headaches that controversial programming inevitably generated. The business of television had consolidated its already dominant position and dispatched a man of ideas. Murrow left CBS, disillusioned and bitter. He accepted President John F. Kennedy's offer to become director of the United States Information Agency.

At Murrow's urging, Friendly stayed on at CBS to continue the work each was committed to. The death of "See It Now" was closely followed by the quiz-show scandals, an event that dropped CBS's prestige to an all-time low.

*Senator McCarthy accepted Murrow's offer of equal time and replied on April 6, 1954.

There were demands for legislative action. Network executives were in a state of near panic. For the first time, the controversy they had tried so hard to avoid presented a real threat to their vital interests. Ironically it was not news and public affairs that brought them grief, rather CBS's entertainment programming. This turn of events caused the CBS establishment to reconsider the public-relations value of news and public-affairs programming. The network desperately needed to rebuild its tarnished image: Public-affairs programming was no longer a luxury; it was a necessity. Thus "CBS Reports" came into being with Fred Friendly as the producer. Despite the intense pressure, CBS was not about to give Friendly the kind of blank check that "See It Now" had enjoyed; those days were gone forever. Friendly would have input and discretion, but not autonomy.

In 1964, Friendly became president of the news division, a position that included, among other things, the authority to preempt regularly scheduled programming for a special news story. The decision to exercise this prerogative was not to be lightly made, since it involved the loss of revenue. Friendly had to make a professional judgment as to whether a news event merited special coverage. If the event was of demonstrable importance, such as a presidential address or a news story of national interest, CBS did not protest. Not to have covered this kind of news would have been a default of its responsibility, thus creating what it most wanted to avoid — controversy. It was the price of running the store, a price that the management of CBS was willing to pay. It was the marginal news story that caused problems — or the important event that was not so perceived by management intent on maximizing profits. The conflict between the business of broadcasting and the public interest became a cause of friction between Friendly and CBS, one that led to his resignation. Friendly's experience illustrates the dilemma; it is a case study on the conflict between corporate interests and broadcasting responsibility.

In February 1966, the Senate Foreign Relations Committee was holding hearings on our escalating involvement in the Vietnam war. An all-star cast of witnesses was to testify: George Kennan, an expert on foreign policy; David Bell, administrator of the Agency for International Development; General James Gavin; General Maxwell Taylor; Secretary of State Dean Rusk. It was the first full-scale national debate on the question of our participation in that war. That it was an important, newsworthy event that merited full coverage was obvious — at least, in Friendly's view. Exercising his journalistic discretion, Friendly covered the first day of testimony with David Bell as the witness. To do so meant that the time normally alloted to daytime soap operas and situation comedies was preempted, with a subsequent loss of revenue to the network. The following day after extracting "reluctant permission" from Frank Stanton, he ordered live coverage of

General Gavin's testimony. At that juncture, CBS called a halt; it was costing too much money, and besides, the housewife audience was uninterested in Vietnam. The following morning when George Kennan was testifying, CBS was showing the fifth rerun of "I Love Lucy." It was an economic decision not a news judgment; the entire news division had recommended live coverage. After appeals to both Stanton and Paley failed, Friendly resigned.

The experience of Edward R. Murrow and Fred W. Friendly in trying to work within the system is instructive. Both were talented professionals who tried to function within a large profit-maximizing corporation without compromising their basic commitment to journalistic integrity. In the end, they had to make a choice between their careers and their principles. To their credit they chose the latter. Their decision had less to do with CBS than with the system that pervades commercial television. Television has a life of its own; the institution dominates the men who purport to run it. Nobody stated this reality more accurately than Mr. Paley in 1936: "Too often the machine runs away with itself — instead of keeping pace with the social needs it was created to serve."[4]

After more than six decades of experience it is apparent that both the rationale and structure of the broadcasting industry are seriously flawed; the machine has run away with itself.

8

Using the System To Beat the System

Reformers within the broadcasting industry are a rarity, and as a rule they do not last very long. Broadcasters must endure criticism from the outside as the price of doing business, but they do not have to take it from the hired hands. In broadcasting, the First Amendment is de facto an owner's right; employees exercise freedom of speech only if they are willing to assume the risks involved.

The threat to broadcasters, such as it is, comes from those institutions having oversight functions: the FCC, the U.S. Congress, and the courts, with only the latter retaining any real measure of independence from the broadcasting lobby. The role of the judiciary is severely limited, its duty being one of enforcing the law. Under the Communications Act of 1934, the FCC has the power to make and enforce rules, impose fines and other penalties; that is to say, to act in both a quasi-legislative and quasi-judicial role. This very considerable power is exercised with restraint, in the full knowledge that many friends of the broadcasting industry in Congress are watching — not the sort of environment conducive to independence by an "independent" regulatory agency.

For the realistic reformer, Congress is not an institution susceptible to pressure; broadcasters are too powerful and legislators too vulnerable. A more innovative approach has been adopted in an effort to force broadcasters to assume their responsibilities to the public: challenging license renewals. This tactic can be effective, but it is also expensive. A full-scale challenge to a license renewal costs several hundred thousand dollars. Obviously, broadcasters are better equipped to finance such a legal battle than are most community groups. Nevertheless, it is a potent weapon in the hands of the public, one that no station owner takes lightly. The mere threat of filing a petition to deny especially from a well-financed group, is a lever that can extract concessions. Few licensees want to throw the dice, no matter what the odds.

The FCC, an agency that has faithfully served the interests of the industry it regulates, would seem to be an unlikely instrument for reform. The community-action groups that sprang up in the 1960s were mindful of this reality, but had little choice in selecting the forum. Procedurally they had to seek relief from the FCC before the case could be heard by the courts. The climate of the times coupled with several new appointments to the commission suggested the possibility that something other than a pro forma hearing followed by a routine denial might result.

In some areas the FCC can be effective. It will take action against a licensee that has violated the commission's rules or provisions of law. Failure to do so would soon result in chaos reminiscent of the early 1920's, a state of affairs that neither the industry nor the public would favor. As the traffic cop of the public airways, the commission does a good job.

On other matters within its jurisdiction, the commission is rarely inspired to take action since to do so would bring it into conflict with the broadcasting establishment and invite congressional pressure, not the sort of thing most bureaucrats welcome. It is easier and infinitely more prudent to avoid controversy whenever possible, to protect the commission budget and staff from congressional wrath, and not incidentally, take a few IOUs from broad- casters that can be cashed in at a later date in the form of job opportunities for ex-commissioners.

Those who attempt to make the system work are not unaware of the realities of power, of the formidable obstacles that will be placed in their path. Despite the elaborate trappings of due process, conducted in a quasi- judicial atmosphere, few are deceived by the elaborate bureaucratic smoke screen. Notwithstanding the FCC's specific mandate to protect the public in- terest, knowledgeable applicants entertain few illusions about the commis- sion and their chances of success.

There are, however, openings through which a determined and innovative advocate can enter, and use commission rules to advantage. More important, the applicant can appeal an unfavorable ruling to the courts. It is a costly, often frustrating, and frequently unrewarding exercise, but occasionally it has proved successful.

In 1967, a young attorney named John Banzhaf wrote a letter to the FCC complaining that television station WCBS in New York had failed to meet its obligations under the Fairness Doctrine by not presenting both sides of a "controversial issue of public importance," namely, the merits of cigarette smoking. The cigarette companies in their advertisements were representing smoking as enjoyable, desirable, and associating it with success and sex ap- peal. There was, Banzhaf argued, a valid opposite view not being presented, one linking cigarette smoking with disease and death. It was, he contended, a "controversial issue of public importance."

Banzhaf's facts and logic were impressive. Numerous private and public studies, including the report of the surgeon general in 1964, had unequivocal- ly stated that there was a health risk in cigarette smoking. What was unusual in Banzhaf's approach was his use of the Fairness Doctrine as an instrument of relief. It had never before been invoked in a product advertising case. Nevertheless, the commission was obliged to rule on it.

The broadcasting industry was more amused than worried. Did this young attorney really think he could successfully challenge the "right" of advertisers and broadcasters to sell products in whatever way they pleased? The FCC could be counted on summarily to dismiss this un-American idea.

The commission with a well earned reputation for avoiding controversy seemed unlikely to rule in Banzhaf's favor. His petition represented a direct assault on the foundation of commercial broadcasting in this country, a system of unchallenged, one-sided, self-serving advertising messages. Banzhaf was trying to change the rules.

The FCC was being asked to break new ground and by so doing antagonize broadcasters. The members of the commission were also aware that each year thousands of people died of lung cancer and other diseases directly related to cigarette smoking. Should it as guardian of the public interest not do something to discharge that responsibility — at least, something symbolic? If keeping people alive and well was not in the public interest, What was?

In employing the Fairness Doctrine, Banzhaf offered the commission some middle ground between doing nothing and total prohibition of cigarette commercials. To the utter amazement of nearly everyone, the FCC ruled that WCBS-TV had not met its responsibility under the Fairness Doctrine, that there was an obligation to offer time to opponents of cigarette advertising.[1]

The ruling ushered in a new era of countercommercials on television warning of the dangers of cigarette smoking, something brand-new in the realm of commercial advertising; a commercial asking people *not* to buy a product. It was an interesting if confusing time with cigarette commercials followed by countercommercials — an utterly ludicrous situation. Only the life-and-death issue tempered the amusement over the advertisers' and broadcasters' discomfiture.

It was the FCC's finest hour. However reluctantly they had risen to the occasion and acted to protect the public interest. Why did they do it? There are several answers, the most obvious being the demonstrable outrage of promoting disease and death on the public airways. The commissioners could not plead ignorance; the evidence of the harmful effects of cigarette smoking was overwhelming.

The second explanation is that the commission merely anticipated events. Congress was under heavy pressure to act and might well pass legislation banning cigarette advertising on television. The mood of the country in the late 1960s was conducive to change with many institutions under attack. Doing business as usual was becoming outmoded.

In making its ruling, the FCC made clear that cigarette advertising was a special case and that it had no intention of setting a precedent for other products of questionable safety. Despite this caveat, it was not long before others sought to employ the Fairness Doctrine. Friends of the Earth complained that WNBC-TV had failed to meet its obligations in not warning about the dangers of air pollution in new automobile commercials.

The commission was not persuaded, refusing to invoke the Fairness Doctrine. Friends of the Earth went to court and won, the Court of Appeals finding the two cases indistinguishable.[2] The result was a settlement between the parties whereby Friends of the Earth received free air time on WNBC-TV.

These developments were alarming. The FCC had intended to open the door a few inches; it was now open a foot. The full wrath of the broadcaster-advertiser lobby descended on the commission with demands that something be done. Broadcasters were furious; they were obliged to give free air time. Advertisers were also upset. The commercials they had crafted with loving care were being contradicted. It was intolerable.

The commission agreed. It corrected its mistake by the simple expedient of reversing the Banzhaf ruling. It was no longer going to assume the "trivial task of balancing two sets of commercials which contribute nothing to public understanding of the underlying issues of how to deal with the problems of air pollution."[3]

The FCC decided to deal with the problem, deemed a "trivial task" by ignoring it, once again permitting one-sided commercials. The public would have to look out for itself. Advertisers, broadcasters, and regulators agreed; the airways were to be used to sell merchandise, not ideas. A brief experiment in broadcasting democracy had come to an end.

9

Censorship: A Little Help from Our Friends

Indeed, in light of the strong interest of broadcasting in maximizing their audience, and therefore their profits, it seems almost naive to expect the majority of broadcasters to produce the variety and controversiality of material necessary to reflect a full spectrum of viewpoints. Stated simply, angry customers are not good customers and, in the commercial world of mass communications, it is simply "bad business" to espouse — or even allow others to espouse — the heterodox or the controversial.

— Dissenting opinion of Mr. Justice Brennan, with Mr. Justice Marshall concurring, 1973

Censorship is antithetical to the American concept of an open and free society. We have chosen instead to have a marketplace of competing ideas with access for many voices diverse and antagonistic to one another, each seeking attention and approval. Out of this process a consensus emerges. It is the American way, and it has served us well.

Those who own or control the mass media agree that censorship is both dangerous and alien to the American tradition. They oppose censorship on philosophical and constitutional grounds, defining it as an act of government that limits or prohibits what can be said, seen, heard, or printed.

Others prefer a broader definition: censorship limits or prohibits what may be said, seen, heard, or printed no matter *under whose auspices this occurs.* Government is not the only institution having a vested interest in manipulation and control. To focus our attention exclusively on government — as mass media interests would have us do — would be a serious mistake. The real issue is whether there is to be a free marketplace of ideas through which the will of the people can find expression. The right to vote is not enough; it becomes meaningful only in an environment of diversity where ideas and information flow freely. To the extent that information and ideas are restricted or controlled either by governmental or private interests, we become a less free people. Censorship fundamentally subverts the democratic process.

Two centuries ago in the era of the penny press when anyone could become a newspaper publisher by investing a few dollars, only government posed a censorship threat. There were no lords of the press or one-newspaper towns. The press was diversified, vital, and contentious. Today in many of our large and medium-sized cities, there is but one newspaper, and this monopoly trend continues. Those papers in existence are the survivors, the winners in the newspaper wars. The real losers are the American people, who are less well informed than they would be were there several newspapers from which to choose.

In fairness it must be said that many of these monopoly papers are of high quality. With no need to compete, some have nevertheless continued to opt for excellence and are among the best in the nation. This does not, of course, argue for or sustain the proposition that to have a monopoly press is good. Quite the contrary, monopoly newspapers are both elitist and dangerous, more consonant with the traditions and values of non-democratic countries. We should take small comfort in the fact that some monopoly newspapers strive for quality, diversity, and a measure of access for its readers. This can all change with the next owner.

There is no easy solution to the problem. Competition and the economics of publishing have left but one survivor. In a society that rewards winning in the marketplace, why shouldn't newspapers reap the rewards of winning? If newspapers were not a special business, there would be little to worry about. However, market forces and good public policy are not always in harmony particularly in the field of mass communications. When newspapers fail society is the real loser.

Unlike newspaper publishers, broadcasters operate in a government regulated marketplace. It is, as a rule, a shared-monopoly environment with spirited competition but not of the kind that is fatal to the participants. Not infrequently, everyone is a winner, particularly in television.

When newspaper owners moved into radio in the 1920s and 1930s, they brought with them many of the ideas and prerogatives associated with publishing. Some found the nominal obligations of public trusteeship burdensome, a governmental intrusion. They wanted to operate "their" radio stations like their newspapers, answerable to nobody. Government regulation was something to be opposed on principle, despite the fact that public trusteeship was essentially a myth — a public-relations exercise. Accustomed to the absolutism of the First Amendment, the lords of the press were being asked to bend a little. It was, many thought, a small price to pay for a lucrative government handout, a subsidy for the rich; but many did not see it that way. They wanted total control.

They have been obliged to settle for less — but not that much less. There are only two obligations that radio and television licensees have that newspaper publishers do not have: an obligation to provide equal opportunities to qualified political candidates in an election, and the obligations imposed by the Fairness Doctrine. Even in these areas, broadcasters have room to maneuver by refusing to sell time to any political candidates and avoiding controversial issues of public importance.

The Equal-opportunities for political candidates and the Fairness Doctrine aside, broadcasters can operate a radio or television station as they please. Licensees control content; it is both their right and duty. They are "designated censors," those who decide what Americans are allowed to hear and see — the gatekeepers of the public airways. Congress, in its wisdom, ceded these powers to the chosen few in the Communications Act of 1934.

Despite the avowed purpose of serving the public interest, no member of the public being served has had any role in policy making. No member of the public as a matter of right serves as a director on the board of any commercial radio or television station in the country. The only people having a voice in policy are stockholders — a very special public. This institutional elitism is a control mechanism that sustains private power. Censorship is a de facto reality in such an arrangement, a governmental grant to a privileged elite. By any reasonable democratic standard of popular sovereignty, our system of broadcasting is a failure. Not only do the few control the many, but that reality has been effectively disguised behind such rhetorical phrases as "public trusteeship" and "serving the public interest." Many are persuaded that there is substance in such impressive declarations; nothing could be further from the truth.

No one who works in or for the electronic mass media has any illusions about where the power lies. Independent producers, film-makers, writers not in the direct employ of the television industry know — and it is their business to know — what is acceptable to their clients, what can be written, said, or filmed. Self-censorship is the rule; overt censorship by broadcasters is the exception. There is little need since those programs that cannot be distributed will not be produced. Freedom in television is quite selective.

Broadcasters are loathe to admit they engage in censorship since only the government can censor. When semantics yield to substance they will concede the obvious; yes, we do make choices; we include some material and exclude other. Somebody must make these decisions.

Clearly some person or group must, but that does not support the proposition that businessmen posing as public trustees are best suited for this task. There are other more democratic alternatives.

10

Public Television: The Poor Relation

When in 1934 the Communications Act was being debated in Congress, an amendment was offered proposing that a quarter of the available frequencies should be reserved for educational and nonprofit broadcasters. As a result of pressure by broadcasting commercial interests, the amendment was defeated, and the matter was referred for disposition to the newly created Federal Communications Commission where it soon perished. The airways were to be used by commercial interests to make money; nothing so inconsequential as education was to be given a share of the electromagnetic spectrum.

This experience was instructive for the educational community, as indeed it should have been for those whose mission it is to teach. They were poorly organized, underfinanced, and politically naive, no match at all for the already powerful commercial broadcasting lobby.

Seventeen years later the educational community returned to Washington, this time demanding its share of the available television channels. Over eight hundred colleges, universities, school systems, and boards of education submitted statements to the FCC advocating such reservation. It was a well-organized, coordinated effort representing a large constituency, one that could not wholly be ignored. The timing was fortuitous, occurring during a "freeze" on license grants brought about by the Korean War and technical problems in the new medium. The emerging commercial television industry was itself disorganized as many contending interests sought the available licenses. Successful applicants acquired wealth and power; the losers like the rest of us would become viewers. In this atmosphere, there was not a lot of cohesiveness, not the kind of unity that might have defeated the educators once again. With hundreds of licenses about to be granted, excluding noncommercial broadcasters was not a matter of high priority; getting one's own "license to print money" was of primary importance. There also was another consideration: when a noncommercial license was issued, this reduced competition in a given market since educational broadcasters could not accept advertising. A noncommercial station was an asset to a commercial broadcasters occupying a channel that might otherwise be in the hands of a competitor.

This marriage of convenience was not a union between equals nor was it so intended. The commercial broadcaster had an economic base; educational broadcasters were beggars living on handouts, an arrangement that suited commercial telecasters just fine. Subsistance TV invited viewers to turn the dial — and they did so by the millions. Educational television (ETV) with infrequent exceptions, was painfully dull, just what commercial broadcasters hoped and expected. It could hardly have been otherwise. While money alone does not assure quality — a self-evident proposition for any perceptive viewer of commercial television — a lack of money makes quality programming next to impossible. Television is enormously expensive; it has an insatiable appetite for money, and no amount of creativity and taste can occupy the void. As a concept, noncommercial television in the United States has never been afforded a realistic opportunity to succeed. From its inception it was doomed, a victim of fiscal malnutrition. Congress, lobbied by commercial TV, has not proved very helpful. It refused to appropriate money or to provide an economic mechanism by which ETV might become self-sufficient.

By 1959, there were only forty-four ETV stations in operation, some broadcasting for only a few hours a day. One hundred and ninety-eight channels reserved for educational broadcasting were unused for lack of money. All this occurred during a period of unprecedented growth in commercial television. While commercial TV prospered, ETV was barely alive.

For the first decade of its existence, ETV languished, unable to attract a large audience or to cast off its dowdy image, a disappointment to its friends, the butt of jokes to its critics. Yet it was not a total disaster. It produced some excellent programs, particularly in the field of public affairs, the sort of programs that commercial television avoids since there was little money to be made from them.

In 1962, the prospects for ETV improved when Congress provided matching funds for educational broadcasting facilities, a welcome stimulus that encouraged many groups to make a commitment. Now, at last, the capital costs were within reach.

Two other events occurred in 1962 that gave momentum to ETV. The National Educational Television and Radio Center moved from Ann Arbor, Michigan, to New York City. With a grant of $6 million from the Ford Foundation, it became the production center for educational TV throughout the nation. It was a large step forward in institutional terms, but there was no network in the strict sense of the word. AT&T lines necessary to provide interconnnection and an operational network were far beyond the financial means available to the affiliates. Nevertheless, having an independent production capacity was an important step forward.

In that same year, a nonprofit corporation purchased Channel 13 in New York for a price in excess of $6 million — a rather generous return on investment by its previous owners. Now at least, the nation's largest city had an educational station, a base of operations to utilize the vast reservoir of creative talent available in what is generally considered the media capital of the world. ETV had emerged from fragile childhood into hopeful adolescence, but the old problem of money remained. It is one thing to have a television station; it is quite another to underwrite an operational budget. Private foundations, grants, and public contributions proved grossly inadequate to support quality programming. Clearly, there was a desperate need for funding on a regular basis. Would the federal government help? It was not the first time the question had been asked, nor was it considered an unreasonable request considering what the government had bestowed on commercial broadcasters. Many other democratic societies have supported public broadcasting for more than two generations.

In 1965, the Carnegie Commission on Educational Television considered the whole range of problems confronting ETV. One recommended change — since adopted — was a change of name; educational television became public television. Among the substantive recommendations was the creation of the Corporation for Public Broadcasting with a fifteen-member board appointed by the President with confirmation by the Senate. A critical recommendation had to do with the creation of a trust fund financed by a tax on television sets. This funding mechanism would provide both money and political insulation — two absolute prerequisites for a truly free mass-communications medium. Congress refused to grant this measure of independence, opting instead for direct appropriations from the federal treasury, a gift given or taken away at the pleasure of Congress. Public television was to be kept on an economic leash, a constraint their commercial counterparts would have considered intolerable. It was something less than an ideal arrangement, but there was no alternative. Pragmatists argued that there was little point in having a medium if it lacked the funds to deliver the message; purists contended it was a capitulation, an abdication of basic principles, one that would end badly.

For better or worse, public television accepted the government dole, and by implication, the inhibitions and proscriptions inherent in a grantor-grantee relationship. Now firmly attached to a congressional umbilical cord, the question became one of relative rather than absolute freedom. The limits were to some extent open-ended; the precise relationship awaited definition. Courage as well as timidity can become a precedent, particularly so when a third force — public opinion — is part of the power equation. Thus it became a test for those who managed public television. How much freedom were they prepared to demand? They could hardly be, or perceived to be, an emasculated tool of government unwilling to take on controversial issues. To adopt this role would have been self-defeating. What is the point of having an alternative network some asked, if it allows itself to become a hostage of the next congressional appropriation? Yes, public television ought to be courageous and deal with controversial topics, but one can only wonder how long or how hard it can bite the hand that feeds it without paying a price. In broadcasting as in any other business, money is power. Without insulated funding, public television can never be as free as it ought to be.

In addition to the not so subtle congressional embrace, public TV must deal with the realities of corporate funding. While commercial advertising is prohibited, PBS and member stations do acknowledge corporate grants. It seems the very least they can do in return for such generosity without which many programs could not be broadcast. This most gentle of soft sells is conducive to feelings of warmth and gratitude by many in the viewing audience, a tasteful change from the cacophony and insult of commercial television, a breath of fresh air that can do nothing but create goodwill for the donor. This is, of course, the intent of such underwriting; the donor is buying good public relations. Grants are not gifts; they are investments, and important ones for corporations that pollute the environment, engage in monopoly practices, or bribe foreign officials. These and other activities create image problems that can affect sales, lead to prosecution by the Justice Department, or unfavorable legislation.

Considering the commercial alternative, it may seem the essence of ingratitude to mention — let alone complain about — the quid pro quo of corporate underwriting, except that such dependence weakens the independence of public television. Not unlike a congressional appropriation, a grant, is subject to review. A congenial association with the donor may be rewarded by more grants. But what would happen if the decision makers in public television broadcast a hardhitting documentary critical of the business practices of its benefactors? To pose the question suggests where the danger lies.

With the creation of the Corporation for Public Broadcasting (CPB), the politicization of public television became inevitable. The board members of CPB are all political appointees, people who have won presidential favor and as a general rule share his political philosophy. Overseeing the workings of public television is, one can safely assume, a matter of more than passing interest to any President.

In 1973, what many had feared and some had predicted became a reality. The board of the Corporation for Public Broadcasting moved to take control over PBS, an affiliate membership corporation that operated the interconnection facilities. Prior to CPB's intervention, PBS had exercised control over what programs were aired on its facilities. It was the media gatekeeper, and predictably, it did not satisfy everyone. Some considered it timid and establishment oriented; others, including the members of the board of CPB, found it permissive and unobjective. CPB decided to correct this situation by appropriating to itself the function of gatekeeper. In a resolution the board of CPB took from PBS "the decision-making process, and ultimate responsibility for decisions, on program production support or acquisition" and "the pre-broadcast acceptance and post-broadcast review of programs to determine strict adherence to objectivity and balance in all programs or series of programs of a controversial nature."[1]

Commenting on this development, the president of PBS, Hartford Gunn, voiced the fears of many: "When you have all the power in the CPB's hands, all the necessary conditions are present for the Corporation to become a propaganda agency."[2]

The rationale for CPB's takeover was an expressed concern for "objectivity and balance," an area in which the board had concluded that PBS was deficient. The remedy was control by fifteen political appointees who better understood "objectivity and balance." With this simple and simplistic declaration the heavy hand of government was at the jugular of public broadcasting.

Observing the disarray in public broadcasting, the Carnegie Corporation funded a second study on the Future of Public Broadcasting in 1977. The study commission consisted of representatives from the arts, business, public-interest groups, and academia. Significantly the commission was relieved of any recommendations made by the first Carnegie study. If this second study was to serve any useful purpose, it would have to be free to seek new solutions to old problems.

In appraising he CPB takeover of PBS, the commission in its 1979 report, *A Public Trust,* stated the following: "Without attempting to judge the motives of the board, we observe that the board took action to downplay public affairs programming in order to avoid placing the entire federal appropriation in jeopardy. Rather than fight for the system's independence from political interference, CPB's decision about NPACT,* various public affairs series, and the takeover of PBS functions seems to us to have been an attempt to mollify the administration in order to maintain *the funding that was now life and death to the system*" (emphasis added).[3]

Having defined the problem as the corrupting power of government money, one might have expected the commission to recommend an insulated funding mechanism (the recommendation of the first Carnegie commission) to correct a fundamental weakness. The commission did consider insulated funding options, one being a tax on the profits of commercial broadcasters. This concept was abandoned as being too difficult to administer. Networks and affiliate station groups "would have a strong incentive to reallocate costs, reducing reported revenue from their profit centers."[4] Broadcasters could shift overhead to reduce profits or change their bookkeeping "in order to hide resources."[5]

*National Public Affairs Center For Television

The commission had little faith in the corporate morality of broadcasters or in the government's capacity to control such activity — and it may have been right on both scores. What it failed to mention was the utter impossibility of ever getting a tax on broadcasters' profits through Congress.

A more interesting idea — similarly rejected — was a tax on commercial advertising. It had the advantage of distributing the burden to many consumers and many industries as contrasted with a more narrowly based tax on television sets (the mechanism used by the BBC and the first Carnegie commission recommendation). The present commission noted that such a tax would be passed along to consumers, thus becoming "not only a source of inflation but a burden on the lowest income levels of the population."[6]

The commission was clearly correct in stating that a tax on advertising would be passed along to consumers, and that such a tax would be inflationary and regressive — precisely what happens now for advertising costs and broadcasters' profits. Implicitly the commission repudiated the current system that sustains commercial broadcasting in the United States.

Study commissions must, if they are to serve any useful purpose, be cognizant of political realities, of what is or is not possible. The commission seemingly decided not to reach for what it could not grasp — an entirely reasonable if uninspired choice. Insulated funding for public broadcasting was not attainable under any formula for the very good reason that advertisers, commercial broadcasters, and their friends in Congress did not want it. Public broadcasting was to be kept poor and dependent, thus more amenable to control and manipulation.

Yielding to the imperatives of entrenched power, the commission decided not to tinker with the system. There was no recommendation for insulated funding, only a plea for more government money. It asked for $590 million in 1985, an increase of $310 million from the 1981 authorization.

11

Children's Television Workshop: A Pearl in the Garbage

We're going to do something like "Laugh-In" for children, with commercials to sell them things they ought to know.

— Joan Ganz Cooney, creator of "Sesame Street"

"Sesame Street" evolved from Joan Ganz Cooney's complaints about the quality of television programs for children. Despite its great wealth and technical resources, commercial television had produced little of value for children. They were content to broadcast cartoons and make money.

A dinner guest of Mrs. Cooney's, Lloyd Morrisett of the Carnegie Foundation, listened to her complaints and the seed of an idea took root: why not produce a quality children's program with the financial backing of the Carnegie Foundation?

A feasibility study was commissioned headed by Joan Cooney, who resigned from her position at New York's public television station WNET. In her formal proposal to the Carnegie Foundation in 1968 she advanced the concept, since adopted by Children's Television Workshop, producer of "Sesame Street," that the techniques of commercial television should be employed. She especially wanted to utilize commercial television's highly successful, attention-getting advertising techniques. As any parent will attest, children are fond of slogans and jingles, testimonials to the adman's craft. The question for Cooney was one of style versus substance, of means and ends. She concluded that this methodology should be embraced even at the risk of offending some cultivated sensibilities. In television as in life, experience is almost always the best teacher. There seemed little point in producing a healthy and wholesome children's program pleasing to teachers and child psychologists; they are not the target audience or the measure of success; the children are. If out of boredom kids tuned out en masse in favor of cartoons or "The Three Stooges," even a superior children's program would be a failure. With the realities of television clearly in mind, Joan Cooney co-opted all the tricks of the hucksters' trade and turned them to her purpose: "To sell them things they ought to know."

Like commercial television's advertisers, "Sesame Street" employed market research as a basic tool, subjecting all theories and program concepts to objective measurement. If the audience — in this case, children in the pre-school age group — disapproved, Children's Television Workshop — took note and tried a new approach. CTW was totally flexible on matters of technique, for as every good teacher knows, there must be a willingness to learn from the student. On substantive goals, i.e., determining what children ought to know — an area of expertise beyond the grasp of the audience — CTW had some definite criteria. All program "bits" — segments running from a few seconds to several minutes — must fall within four categories: symbolic representation (letters, numbers, geometric shapes, and key words); cognitive organization (matching and discovering shapes of objects, classification of objects by size, shape, or function, and the like); reasoning and problem solving (drawing inferrences, predicting consequences, making evaluations and explanations); the child and his or her world (understanding social relationships, etc.) Each program has a theme, a precise plan intended to accomplish certain objectives — a number or letter, a geometric shape, a social concept, a sound, a relationship, etc. All programming for "Sesame Street" is pretested by a group of educational psychologists; nothing is left to chance or unevaluated theory.

Measuring the level of interest in children ranging from two to five years in age whose language skills are still not fully developed, presents certain problems. To frame appropriate questions and obtain useful answers is not easy. To work around this problem, an ingenious "distractor" test has been employed. It involves a slide projector with an automatic carousel aimed at a screen located adjacent to a television set. At periodic intervals, pictures of animals or cartoon characters flash on the screen, in effect, competing for the attention of the child. If what is being shown on television fails to interest the child and hold his or her attention, there is an alternative available. Thus if some part of "Sesame Street" loses the attention of too much of its audience, it is changed or abandoned. Responding to the viewing interests of children is a guiding principle, for learning and interest are closely related. The challenge of "Sesame Street" is teaching within a framework of entertainment, a "fun-and-games" approach that some educational traditionalists regard with suspicion, even hostility. Within a structured school environment, where regulation and control are viewed as prerequisites to learning, fun and/or games have a low priority. Not all educators accept this artificial dichotomy, however; they see no reason why learning cannot and should not be enjoyable. Besides, they argue, such an approach works. Given the fact that traditional methods of teaching seem to be failing, often spectacularly, perhaps more emphasis should be placed on new, innovative techniques designed to inspire and motivate, rather than to coerce. The most notable achievement of "Sesame Street" may be its stimulation of childrens' innate intellectual curiosity, a love of learning. If he or she is fortunate enough to encounter the right kind of teachers later on, this gift may last a lifetime.

In devising the format for "Sesame Street," Mrs. Cooney decided to use cartoons for the very good reason that kids loved them, even if some child psychologists and educators frowned on them. It was another example of favoring ends rather than means. Cartoons per se were neither good nor bad; it was the content that mattered, not the art form. So long as they were her cartoons, selling her message, there was no objection from Joan Cooney. An additional advantage was cost — particularly in presenting pixilated cartoons, a herky-jerky motion created by omitting connecting frames, that the children liked. It produced fast-paced, exciting action, at a fraction of the cost of conventional cartoons.

Simplicity and repetition are basic "Sesame Street" themes. It is very difficult for a child to learn and remember, no matter how easy it may seem to an adult; it requires many different explanations, repeated many times before it is completely understood and retained. To reinforce this process, "Sesame Street" programs are repeated often, and the audience does not object. In fact, many children are delighted by their familiarity with the characters and content of previous programs, affording them the opportunity of meeting old friends, as well as demonstrating how much they have learned.

"Sesame Steet," like its commercial competitors, has adopted a star system, a cast of familiar performers with whom children can identify. This identity bond, it was discovered, enhances the capacity of the character to teach as well as to entertain. Thus Big Bird is not only the largest and most lovable bird extant; he is also a friend and confidant of millions of attentive viewers. When he turns philosopher in a morality play, his friends watch and listen respectfully.

The level of sophistication on "Sesame Street" is high. There is no talking down to the audience. Positive social values are inferrentially stressed; there is no hard sell. Moral declarations, however worthy, frequently do not register as abstractions, but in the simulated experience of "Sesame Street," such concepts assume meaning.

One of the major achievements of "Sesame Street," and perhaps the most enduring, is in the area of human relationships. By teaching positive social values and acquainting children with a diverse pluralistic world, filled with many different people all sharing a common humanity, "Sesame Street" is often inspirational. Academic shortcomings are remediable; social malperceptions acquired at an early age can persist for a lifetime.

"Sesame Street" is concerned with the whole child, rather than with some limited educational objective. It has received an impressive list of awards for its many achievements, but the best testimonial is its loyal audience of millions who regularly watch this marvelous program.

The success of "Sesame Street" offers a challenge to commercial broadcasters, demonstrating what can be accomplished given adequate financial and creative resources. Commercial broadcasters could, if they chose, produce the equivalent of a "Sesame Street" — or even several such programs.

What they lack is the incentive. Why aspire to quality when the kids are content with cartoons?

12

Concentration: Television's Favorite Game

In considering the public interest the Commission is well within the law when, in choosing between two applications, it attaches significance to the fact that one, in contrast to the other, is disassociated from existing media of mass communications in the area affected.

— Scripps Howard Radio, Inc., FCC, 1951.

It was predictable, and perhaps inevitable, that the patterns of ownership established in radio would continue in television. Given the political realities, no other result was remotely possible. Large, well-established media interests understood that television represented the future in mass communications, and they wanted as large a share as possible. The logic was compelling: if one major market television station was a potential gold mine, why not own many stations?

In theory, the FCC favored the principles of diversity and local ownership, but when it came time to award licenses, these considerations were quickly abandoned under heavy pressure from media interests and their friends in Congress. Ultimately the FCC allowed a single entity to hold seven television licenses, five of which could be in the VHF bandwidth.

Few would argue with the proposition that a free society is best served when information and opinion emanate from a variety of sources, independent and in vigorous competition with one another, allowing the reader, listener, or viewer to make judgments predicated on a broad spectrum of facts and opinions. This belief, which was central to the thinking of the Founding Fathers, is given expression in the First Amendment. Only in such an environment of openness and diversity can individual freedom and democracy survive.

With the passage of time and the advent of technology, the street-corner orator has given way to the overwhelming influence of radio and television; the penny press has been replaced by monopoly newspapers.

To examine the dimensions of concentration in television, a medium in which diversity could be a matter of public policy, the author surveyed the ownership of TV stations in the twenty largest markets.* These markets are especially important for a variety of reasons. They are located in our most populous states, where more than one-third of our people live. All fifteen

---

*Data collection *Television Factbook, 1981-1982,* services and stations volume, Washington, D.C.: Television Digest, Inc.

network-owned and -operated stations are in the twenty largest markets as are the network production facilities. What happens — or does not happen — in these markets affects every other market in the country.

Noncommercial stations have not been included, nor have UHF (ultrahigh frequency) stations, none of the latter having a network affiliation. However, independent VHF stations have been included since several exert considerable influence in their respective markets.

Ownership of VHF stations in the twenty largest markets is a complex maze of communications-related businesses, but for the purposes of this survey only four have been considered: Network-owned and -operated stations, group-owned stations (from two to five), stations owned by newspapers, and stations owned by newspapers publishing in the same market.

There are seventy-seven VHF stations operating in the twenty largest markets. Of these, fifteen are network-owned and -operated; one is a CBC affiliate, CBET, serving the Detroit market; forty-five have a network affiliation; and sixteen are independent. Only six stations are owned by interests having no other TV holdings. Newspapers own, in whole or in part, interest in twenty-eight stations; fourteen publish and broadcast in the same market, a particularly unhealthy situation. Illustrative of this is the Dallas-Fort Worth market with three of the four VHF stations owned by newspapers, two publishing in the same market. The Houston market is yet another egregious example. Two of the three VHF stations are owned by newspapers, one publishing in the same market. The consequences of such cross-media ownership are difficult to measure, but it is worth noting that as a percentage of revenue the television profits in the Dallas-Fort Worth and Houston markets are among the highest in the industry.

In such an environment the political process can be effectively compromised, manipulated, or corrupted by the exercise of private power. What politicians would willingly choose to do battle with these vested interests?

Clearly, the most dangerous concentration exists in the five largest markets: New York, Los Angeles, Chicago, Philadelphia, and San Francisco.* In these markets there are twenty-four VHF stations, nineteen owned by only seven interests. They are: CBS, NBC, ABC, Metromedia, *Chicago Tribune–New York Daily News,* RKO and Group W.

The importance of these stations in political terms can hardly be overstated; their influence is enormous, a reality that is a given for every presidential candidate. These are the media nerve centers of the nation, located in our most populous states. No candidate can expect to be nominated, let alone elected, without doing well in these states.

*Arbitron ADI ranking, *Television Factbook, 1981–1982,* services volume.

The three networks own and operate eleven stations in the five largest markets and have a monopoly on network programming in the three largest markets. One can argue whether this power is being used wisely or well; the point is that it exists. One need not subscribe to the conspiracy theory of history to have misgivings concerning the wisdom of having so much power in so few hands, people elected only by their stockholders. Ironically, those who acquire power by facing the electorate must first pass through the filter of the electronic media, a control mechanism more powerful than the politicians it scrutinizes.

Sophisticated politicians understand television. They understand what the medium can do for them and to them; they know the rules of the game and how to play it. Those who do not become media nonpersons. Creating media events and uttering quotable quotes in time for the evening news — software for the electronic gristmill — are requisites for a performer-candidate.

Politicians use television, and television uses politicians, particularly those seeking election to federal office. By reason of his need for favorable television exposure, a politician can ill afford to take positions or advocate legislation hostile to the interests of station owners. These interests are not the narrow concerns of the medium, which are sacrosanct in any event; rather they are the broader interests of conglomerate ownership. Politicians with an instinct for survival are not unaware of who owns what.

The power of television is so pervasive, so demonstrable, that it need not be used indiscriminately. When employed to punish, it can appear as an exercise in objectivity to the casual observer. The scalpel is used, never the headsman's axe; there is no bleeding, no claim of victory. The passion and gloating of the old-time press lords' vendettas are mostly absent in television for to declare a victory is to admit that there was a victim, something an objective medium like television eschews. When someone is defeated, he must by television's rules be a victim of the facts or of himself. The camera never lies.

13

Cable Television: The Third Wire

We are in the process of a communications revolution. To predict what a revolution will bring is not easy. To attempt to write about the process while it is still aborning undoubtedly means that much that we write today will look foolish in five or ten years' time. One is tempted to sit back, wait, and see what happens. But if we wish to influence the outcome rather than just observe it *post hoc,* we must do our best to understand it better while it is still in process.
— Ithiel de Sola Pool, *Talking Back: Citizen Feedback and Cable Technology,* 1973

Cable television came into being as a result of the technical limitations of broadcast television, a line-of-sight medium unable to pass through hills and mountains. As a consequence, those people who lived in remote areas or mountain valleys had either poor television reception or none at all. The golden medium was unavailable to millions of Americans.

Into this first television wasteland there came an assortment of entrepreneurs ready and willing — for a fee — to provide a good television signal for their subscribers, giving reality to the American dream — a television set in every living room.

The first cable television system was built in 1949 in Lansford, Pennsylvania, by an appliance-store owner.* He was convinced that the sales of television sets would increase if people could get good reception. This gave birth to an idea: why not provide that reception? To do so did not involve creating a new technology; the principles and hardware for a commercial cable system already existed.

The formula was simple: find a community with poor television reception or no reception; go to the top of the highest mountain or hill and build an antenna that could receive over-the-air television signals; amplify the signals and transmit them to a headend (the cable equivalent of a broadcaster's transmitter), then distribute those signals by coaxial cable to subscribers. The result was a good picture and a variety of television signals.

A new service had been created, one that resulted in a love-hate relationship between broadcast television and cable television. The former regarded the latter as a "siphoner," one who used a broadcaster's signal without paying for it. Cable operators argued that by using the broadcaster's signal he expanded the broadcaster's audience, and consequently his profits. It was a marriage consumated in mutual self-interest; everyone prospered.

*At about the same time another cable system was constructed in Oregon.

This grudging co-existence was severely tested when a cable operator decided to expand the service provided to his subscribers by importing distant television signals via microwave. By so doing, subscribers has more diversified programming, more choices.

Television broadcasters were outraged. Competition from stations outside their market fractionalized the audience; it was bad for business. They argued that competition would lead to bankruptcy. Television broadcasters appealed to the FCC for relief. To no one's surprise, the FCC subsequently ordered a freeze on the importation of distant signals into the one hundred largest television markets.

It was a victory for broadcast television, but it was also a warning. These upstart cable systems though temporarily constrained by regulations imposed by the FCC had the technical capacity to offer serious competition. Television station owners were worried, mindful of radio as a precedent. Radio had been the dominant electronic medium until television came along, and some observers believed history might repeat itself.

These fears have escalated in recent years with the entry of large, well-financed communications conglomerates into cable television, obviously persuaded that they were buying the future. Many of these companies have had vast experiences in both print and electronic mass communications, some with substantial holdings in radio, broadcast television, newspapers, magazines, book publishing, and program production. A partial list of the companies with heavy investment in cable television is, at the very least, suggestive: Time Incorporated, the New York Times, Los Angeles Times, Chicago Tribune, Dallas Times Herald, San Francisco Chronicle, Cox, Newhouse, Westinghouse, UA-Columbia, Warner Communications.

If these experienced and successful companies now invest hundreds of millions in CATV rather than some other communications medium, it would be useful to know why, to ask what plans they have for the future, both for themselves and for the general public. Obviously, these companies believe the prospects for cable are bright, that there is money to be made.

Most people think of cable in terms of receiving more channels, more programming choices, more entertainment. Cable-system owners have a different view, a long-term view. Yes, the home-entertainment market for quality programming — for pay cable — is huge and barely tapped at the moment despite a spectacular record of growth. There is and will continue to be a market for a variety of home entertainment on a subscription or pay-per-view basis. But the basic and pay-cable services will produce only one-third of cable revenue by 1990 in the opinion of Howard Anderson, president of a marketing and research firm. One-third will be generated by data transmission and one-third by transactional services (banking, merchandising, etc.).[1]

Even if these projections are wrong — and some would argue that they are too optimistic — a central fact remains: cable is more than an entertainment medium; it is or soon will become a *communications system.* The distinction is all important. Unfortunately, the entertainment potential of CATV has obscured other, larger, and more profound applications of this emerging technology. The public-policy debate on cable has focused on what it is now, rather than what it may become; too little thought and even less planning has been given to the future, a state of affairs not displeasing to cable operators.

The FCC also views cable as a communications system, not simply an entertainment medium: "We envision a future for cable in which the principle services, channel uses, and potential sources of income will be from other than over-the-air signals."[2]

Cable, the third wire into the home, like telephone and electric service, has evolved in a monopoly environment. Currently, fewer than ten cable companies compete in the same geographic area despite the nonexclusivity clause in most franchising agreements.[3] It makes no more economic sense for cable companies to compete than for telephone and electric companies.

For the first quarter of a century of its existence, cable was little more than an extension of broadcast television. There was some local programming — sports, news, city council meetings, and public access — but many systems having twelve or fewer channels did nothing more than retransmit over-the-air signals. Origination programming was expensive, a drain on profits. Most subscribers were content receiving a good signal and a variety of stations.

Offering only what most Americans got "free," cable grew slowly. In 1963, there were only 950,000 subscribers in the country. By 1973, there were 7,300,000.[4] Cable was a success with a solid but hardly spectacular record of growth. Broadcast television had all the glamour and most of the money. Cable was still an appendage, a relay service.

In the middle of the 1970s, the perception of cable began to change. One event in particular attracted attention. On September 30, 1975, Home Box Office (HBO), a pay cable service owned by Time Incorporated, relayed via satellite the "Thriller From Manila," a heavyweight championship boxing match between Muhammad Ali and Joe Frazier. In April of that year, Time had invested $6,500,000 for a five-year lease on transponder time on RCA's Satcom One satellite. The purpose was to make Home Box Office a major program supplier for cable systems. Most cable operators were dubious about HBO's prospects, and only two systems agreed to carry the Ali-Frazier match. It was a technical achievement but a market failure.

One interested observer was Ted Turner, flamboyant sportsman and owner of television station WTCG (now WTBS), a nonaffiliated UHF station in Atlanta, Georgia. Seeing what Home Box Office accomplished gave him an idea. As owner of the Atlanta Braves baseball team, the Atlanta Hawks basketball team, plus a large film library, he had something to sell to cable systems scattered around the country; the problem was one of distribution. Conventional distribution via landlines was cost-related to distance, thus economically unfeasible. Satellite transmissions were not. It cost no more to relay a signal from New York to Chicago than from New York to Los Angeles. Although nobody had yet tried this, Turner was convinced it would succeed. Cable systems all over the nation with excess channel capacity were anxious to provide more diversified programming for their subscribers. Turner's money-losing UHF station, a minor force even in the Atlanta market, became a highly successful "superstation" far outdistancing several late-arriving competitors.[5] Home Box Office has been equally successful. It is the largest pay cable program service, more than twice the size of its closest competitor.[6]

Ted Turner and Home Box Office proved their thesis that if you have something to sell (programming), and a means of cost-effective distribution (satellites), there was a market ready and waiting. And it was a huge market. People wanted quality and diversity; they were tired of the lowest-common-denominator offerings of the commercial networks and were willing to pay for something better.

Unfortunately for Turner and HBO, most of the market was foreclosed because the majority of American homes lacked cable service. At the end of 1979, 79 percent of American homes did not have cable.[7]

The solution was obvious: make cable available to those without it. This was the vision expounded by Ralph Lee Smith in his book. *The Wired Nation* written in 1972.[8] His prescient and seminal exposition of cable's potential, dismissed as futurology or science fiction by some, is rapidly becoming reality. The events of the past few years are truly remarkable. Currently forty-nine of the fifty largest cities are in some stage of cable development.[9] By the end of the decade an estimated 58,900,000 homes will have cable , 62 percent of all TV households. If anyone had suggested such numbers in the mid-seventies, most cable operators would have scoffed at the idea. Although wiring the urban centers had always been their objective, many believed it might not happen for years, perhaps not in this century. Cities with few exceptions had good reception, three network affiliated stations, several independents, and a PBS affiliate — enough variety for most people went the reasoning. Besides, who wants to pay for programming? Isn't television supposed to be free?

Suddenly, this conventional wisdom was questioned by the experience of pay cable. Rural and suburban subscribers had a variety of over-the-air signals — in many cases more than their urban neighbors — yet pay cable had been successfully marketed, with more than half (54.29 percent) of the cable households subscribing to pay cable.[10] There was no reason to believe that urban cable subscribers would be any less interested in quality programming than suburban and rural subscribers. In fact, the demographics for many urban areas suggested that pay cable would win wide acceptance.

The rush to wire major urban centers began in the late 1970s and became a stampede in the early 1980s. Cities were besieged by cable companies ready — indeed anxious — to invest hundreds of millions of dollars. Urban cable had become a glamorous growth industry. As in the early days of broadcast television when licenses were being given away, getting in early was all important. Franchises for cable are not bestowed forever — typically for fifteen years — but like television licenses they tend to be forever. Once in, cable companies are able to exert strong leverage to stay in.

Bankers were as enthusiastic for urban cable as the cities were to get it. Despite a sluggish economy and high interest rates, cable companies were able to borrow heavily to meet the large capital costs. Amortization of these costs was projected over a seven- to ten-year period.

At the current rate of construction, most cities in the continental United States will be fully wired by 1986. The third and most important electronic mass communications revolution will have begun.

14

QUBE: Cable Coming of Age

If through some technical limitation the telephone was a one-way medium where one could listen but not reply, its utility would be, of course, markedly reduced. It might still be useful — as letters and telegrams are useful — but hardly the essential communication tool that it is. The telephone is valuable because it is an interactive medium.

The more flexible a communications system is, the greater variety of modes it can accommodate, the more useful it becomes. The Bell telephone system understood this relationship and asked a question: Would our customers want and be willing to pay to see as well as to hear the person with whom they were talking? It was, Bell believed, a logical extension of an existing system. It was logical; it was also wrong — the Bell equivalent of the Edsel. The coupon-sized, low-resolution picture failed to interest people, and the cost was high. Many considered it a gadget. Bell withdrew the picturephone from the market but did not abandon the idea. In the past few years the concept has been reborn and given a new name: teleconferencing, an interactive, audio-video communications system that has proven highly successful.

Broadcast television despite its pervasive influence and technical excellence is a one-way medium. It can never be a *communications system*. Cable, on the other hand, has the capacity to carry all the electronic media — radio, broadcast television, digital, and telephone service — on a single wire. It can be interactive in all modes: audio, video, and digital.

To say that cable has a technical capacity is one thing; it is quite another to predict when, if ever, this potential will be utilized. The marketplace is not a slave to technology. A better mousetrap must not only be better; it must be affordable. The economics of a fully switched, on-demand audio-video-digital communications system are well out of the reach of the average person. For the time being, audio-video interactive communications are restricted by cost for special needs — medicine, business, etc. In time, the economics of scale should reduce the unit costs of this expensive hardware, but something approaching the experience in electronic calculators — from several hundred dollars per unit to less than ten currently — is needed before an all-mode interactive cable system for general use becomes feasible. Costs aside, the inherent technical capacity of cable is considerable, expandable both quantitatively and qualitatively.

At present, interactive cable is commercially viable in the digital mode with the first system called QUBE, having been in operation since 1977. Built by Warner Communications (now Warner-Amex), QUBE's Columbus, Ohio, system has served as the prototype for other interactive systems (Cincinnati, Houston, Pittsburgh, and five other cities that will have QUBE technology).

In the mid-seventies when other multiple-systems operators were being cautious, Warner invested some ten to twenty million dollars — estimated cost since Warner has declined to reveal the precise amount — for wiring Columbus with an ultramodern cable system, interactive in the digital mode. With home terminals linked to computers, information can be sent "upstream" as well as "downstream." Cable television thus became a communications system; subscribers could interact, talk back to their television set. It was a seminal event in electronic mass communications for which Warner Communications received a great deal of attention and praise, much of it deserved. Warner took a chance on cable and won.

Currently Warner has more than fifty-eight thousand subscribers in Columbus, 70 percent of whom have opted for QUBE service — a very high ratio. The reasons for QUBE's popularity are many: the quality and variety of programming — movies, sports events, entertainment specials, local programming, etc.) — but the big attraction is the system itself. Subscribers play with QUBE, interact with it. Unlike one-way cable or broadcast television in which the viewer is relegated to passivity, QUBE subscribers have a role to play. For example, QUBE subscribers named a newborn baby from a list of choices offered on their television screens, accomplished in the comfort of their living rooms merely by touching a button.

QUBE offers its version of the "Gong Show" where performers found wanting in talent are given a collective electronic hook. What power! — and a lot of fun, too.

There are other more substantive applications: one can order a championship boxing match or other sporting event on a pay-per-view basis, can see an uncut movie without commercials, or buy products — including books — by the press of a button. QUBE markets a "duress" button with which to summon police or medical assistance. It also offers a fire-alarm system wired to the local fire station. Upon receiving an alarm, the fire department will respond to the call knowing how many people live in the house, whether there are any inflammables in the vicinity, and the location of the nearest hydrant, all on a computer printout. And this is only the beginning of a dialogue, a first-generation two-way cable system.

As with many advances in technology, there are unfortunately attendant problems. The same communications capacity that allows a subscriber to buy a book or pay for a movie can collect information on one's viewing or reading habits, store it for future use, and/or disseminate it to others for use or misuse. Do we really want our personal tastes recorded or our purchases in somebody's computer bank, to end up on yet another mailing list? And suppose you read a particular kind of book or watch a certain type of movie, should that be a matter of public information for others to use as they please? Most would shout an emphatic no to such invasions of privacy, yet that is the potential of this communications system.

Although Warner-Amex has scrupulously guarded this information mindful of its explosive potential, it is but one company's response. Policy can change. What is needed is legislation at the federal level prohibiting the selling or exchange of information without the approval of the subscriber. And the penalties should be meaningful: a jail sentence for the offender and the loss of a cable franchise for the corporation. These sanctions, if enforced, would go a long way toward solving the problem. Corporate officers would guard confidential information with the same enthusiasm they guard their money.

Another worrisome problem is electronic voting. It is certainly far more convenient to touch a button on a home terminal than to wait in line at the local schoolhouse. Proponents argue that it would encourage broader citizen participation in the political process. With presidential elections decided by slightly more than 50 percent of the eligible voters and many local contests decided by a minority, it is appropriate to employ two-way cable. Each voter could be given a personal identification number (PIN), the same system employed in electronic banking. With suitable safeguards, a workable, cost-effective system could be devised.

Critics are far less certain. Even if a safe system could be devised, some are not persuaded that it would be a good idea. It would reward the lazy, ill-informed voter, one who neither knows nor much care about the issues or candidates. While there is something to be said for broader participation in the political process, electronic voting is not the answer; it is, in fact, a prescription for chaos. Inevitably people will want to vote on everything, each city-council ordinance, each county or state legislative act. It seems wiser to stay with the present system.

Although we are some years removed from making a decision about the merits of electronic voting, it is not too soon to begin thinking about it. Far too often technology overtakes and overwhelms us, creating new realities with little or no debate about the public-policy ramifications. Our experience with broadcast television is an illustration of this phenomenon with its reshaping of our society, our habits and values. The medium is now the master of its own destiny, beyond effective social control.

Now the third generation in electronic mass communications has come to pass and with it new challenges. Unfortunately the opportunity that cable now provides has obscured other issues. Cable for most people suggests entertainment, more and better programming choices; few think of it as a communications system, and therein lies the problem. Cable is being marketed as an entertainment medium; in consequence, some profoundly important questions have not been asked. We have no overall communications policy, and we badly need one — sooner rather than later given our experience with broadcast television. Once their power is consolidated, cable interests may be difficult to dislodge. If we passively wait to see how cable evolves from an entertainment medium to a broad-based communications system, we may become the prisoner rather than the master of this technology; many of the choices may be made for us. We should not be sanguine about popular sovereignty; it is a principle much honored in the breach.

As cable moves into the field of so-called "enhanced services," many policy questions remain unanswered. As a de facto monopoly cable merits attention. Currently, only basic cable services are regulated. All other services, including entertainment options, are unregulated. The cable operator can charge whatever he can get. That may be all right for movies and sporting events, but suppose one wanted some of the following services: electronic banking, computer access or burglary and fire alarm systems? (To mention just a few.) Should the cable operator be able to charge what he pleases? whatever his position in a monopoly market allows? Yes, we may muddle through and come out with a sound public policy; then again we may not. It is useful to remember what happened to the public airways we own.

15

Cable and Education

With the emergence of cable as a major communications system, some educators are delighted, believing that it will play an important role in education. Others are not so delighted or so optimistic, viewing cable as a threat. Oviously, if a classroom teacher with thirty students can be replaced by an electronic teacher with three hundred or three thousand students, many teaching positions will be eliminated. It is basic arithmetic, the economics of scale applied to education.

Understandably it is difficult for all educators to view televised instruction with total objectivity; many have a strong interest in the status quo. The conventional view is that television is a tool, an adjunct to the educational process. It will never replace the classroom teacher. Some regard television as the problem — not the solution — a passive medium that stifles intellectual growth.

The use of television in education neccessitates the expenditure of money by the local school board and the cooperation of administrators and teachers. As in most institutions, change comes slowly. Administrators err on the side of caution; the tendency is to proceed slowly, carefully. Administrators are reluctant to make fundamental changes, thereby upsetting powerful interest groups. Teachers and their unions are a force to be reckoned with, particularly in large urban areas. It is easier to get along by going along. In consequence, television has played a relatively minor role in most school systems.

Politics aside, the question remains: Can television teach? If we define teaching broadly, the imparting of information from one to another, television is an accomplished teacher. It teaches small children about the world that they will soon inherit; it defines our values, our competetiveness, our proclivity for violent solutions. Yes, television can teach, far better than most people suspect. We all learn from television, children and adults alike. Save only for the printing press, no other technological innovation has had so profound an influence.

Viewed narrowly, and from a pedagogical perspective, some ask how effective — and cost-effective — is television. Can it replace the classroom teacher or is it merely a supplement, an "enrichment" option, the frosting on the educational cake? Enrichment television is not intended as a substitute for the classroom teacher, nor is it intended to reduce educational costs. Quite the contrary, enrichment television adds additional costs. When enrichment television is offered, classroom teachers are not economically threatened. They will not lose their jobs no matter how good the instruction is or how they compare to their electronic counterparts. They are free to praise or condemn, to make use of the content or to ignore it. The superior teacher who feels secure about her or his ability may welcome the experience; the inferior teacher may well resent the inevitable comparison.

Instructional television (ITV) is quite another matter since it involves both professional sensibilities *and* economic self-interest. The intent of ITV is, among other objectives, the saving of money by eliminating teaching positions, a goal not designed to win universal approbation from teachers or their unions. ITV is to education, what Henry Ford was to the automobile. The craft of imparting knowledge, like the craft of building the automobile cannot but undergo a radical transformation with the introduction of technology. Much as we may yearn for the return to a simpler life or deplore the often dehumanizing aspects of mass production, the fact remains that few of us could afford a handmade car even if we wanted one. It may be that we can no longer afford to educate our children in the way our great-grandfathers were. Perhaps it is time even for education to embrace technology — or, at the very least, examine the idea with an open mind. If ITV can do the job, or part of the job, of educating and at the same time reduce costs, it ought to be given the opportunity. Viewed in the content of rising educational expenditures, the tax burden is becoming unbearable, particularly in large cities with a diminishing tax base. Obviously, any savings that could be realized would be welcome if, in the process, the quality of education does not suffer. The proponents of ITV claim that this result is not only possible; it has in fact been realized in many schools and colleges that have given a meaningful commitment to it. Token efforts employing poorly trained teachers unfamiliar with the demands of the medium, an inadequately designed curriculum, and poor quality hardware will produce high-cost, unsatisfactory results. To be effective — as well as cost-effective — ITV must have high quality in teachers, curriculum, and hardware. In addition, there must be a large student base because per-student cost is related to the number of those participating. When the scale is on the low side, per-student cost rises, making ITV uneconomic.

Public education in the United States is by far the largest budget item for cities and towns, often consuming more than 50 percent of the tax dollar. Salaries constitute from 65 to 75 percent of educational costs. If savings are to be made on any significant level, they must come from salary expenditures, precisely the area in which ITV can play an important role. In the absence of innovation and technology, educational costs seem destined to rise. Nobody wants to shortchange our children; a good education is — or ought to be — an American birthright. In recent years, taxpayers in many school districts have been in revolt over escalating educational budgets, the general perception being that they are paying a great deal more for education and getting a lot less in return: high school graduates reading at grade school level, unable to write a simple declarative sentence or do basic arithmetic. There are many explanations for these results; still the problems persist. Taxpayers are less inclined than ever to give educators the latitude and decision-making powers they once enjoyed. The failures have raised some fundamental questions: Are educators sufficiently flexible and innovative? Can change come from within, or must it be imposed from without? The era of deference has passed; the people who pay taxes want results, not another educational theory.

Not all educators resist change; many are acutely aware of the problems and want to solve them but are constrained by institutional inertia. As in any bureaucracy, change occurs incrementally; innovators are viewed with suspicion, reformers with hostility. The most creative ideas often fail to excite a comfortable, entrenched bureaucracy. Advocates for change soon become discouraged and retreat into silence.

Given the right environment, innovative ideas do emerge, survive, even flourish. Such a place was Hagerstown, Maryland, where in 1956 an educational experiment began. With a grant from the Fund for the Advancement of Education, Hagerstown in association with the board of education of Washington County, initiated a five-year experimental study in the use of closed-circuit television (CCTV). Preliminary results were encouraging, and the program was retained and expanded. Now more than a quarter century later, the marriage between technology and education continues. Hagerstown has the largest and most comprehensive closed-circuit educational television system in the nation with forty-five schools linked by coaxial cable. Programs emanate from a central studio employing six channels.

The cost of building and operating the CCTV system during the test period (1956–1961) totaled $1,300,000. Commenting on the cost of employing CCTV, the *Washington County Closed-Circuit Television Report* declared: "There was a time when it was thought that the cost of a county closed-circuit network would make its use prohibitive. But this has not been the case. The redevelopment of personnel and equipment made possible by television has produced savings which cover the annual operating costs. And in terms of duplicating in conventional classrooms what is now offered on television, the county's savings are substantial. Without television, the county would require more than one hundred additional teachers and a budget increase of almost $1,000,000 to duplicate the courses that have been added to the instructional program. This is more than three times the annual operating costs of the television network. For example, without television it would cost more than $250,000 annually to provide art and music specialists for the elementary schools."[1]

In a study done for the U.S. Department of Health, Education and Welfare by the General Learning Corporation, the cost for 10 percent instructional television time was thirty-three dollars per student.[2] The cost for Hagerstown was sixteen dollars per student for 15 percent instructional television time.

The Anaheim, California, school system has heavily committed itself to ITV, also employing a closed-circuit system. In 1959, an ITV system became operational using a "redeployment plan," a system with large classrooms for television instruction in a variety of subjects: science, art, music, social studies. These rooms held as many as seventy-five students. A skills classroom with a student-teacher ratio of 25 to 1 was used for more intensive instruction in such subjects as languages, reading, and math.

As a result of the savings realized by televised instruction, class sizes could be reduced for other studies. By relieving the workload for classroom teachers, time was made available for other duties.

The Anaheim experience was similar to that of Hagerstown in terms of cost effectiveness. Its per-pupil cost was lower than the state average and lower than other school districts in the same county for the first three years of operation.

Educational results were equally satisfying. The overwhelming majority of students learned as well or better with the use of television as they had with conventional instruction.

This is an experience typical of ITV. In 393 ITV programs, results were compared to conventional instruction with the conclusion that "in 65% — there is no significant difference. In 21%, students learned significantly more (by television); in 14%, they learned significantly less from television."[3]

Evaluating the Hagerstown results, the Rand Corporation correlated learning achievement in science by teaching method and IQ. The findings are illuminating:

Students Taught Conventionally		Students Taught by Television	
Average IQ	Growth (in months)	Average IQ	Growth (in months)
117	12	118	15
100	11	100	14
83	6	83	13

ITV data strongly suggests that students can learn by the use of television. Still, there is widespread skepticism and no small amount of apathy in the educational community.

Some legitimate concerns about cost are raised. Building a quality, state-of-the-art ITV, closed-circuit system is expensive. Despite the encouraging experience of Anaheim, Hagerstown, and other communities, school boards are reluctant to commit the substantial funds, particularly when confronted by determined opposition from teachers and administrators. Politically ITV is a risky proposition. If it fails, for whatever reason, proponants will take the blame.

A dedicated ITV, closed-circuit system has one major drawback: for most of the time the very expensive hardware will be idle or underused. A dedicated system by its very nature involves a measure of inefficiency, a state of affairs not unlike existing educational plants where schools lie idle on weekends, vacations, and summers. A business could hardly afford the high capital costs and low utilization; only the taxpayers with little choice in the matter have to endure such inefficiencies. The question becomes one of relative cost and utilization. If, however, ITV were made part of a *multipurpose* cable system, the costs would be markedly reduced. A dedicated ITV system involves different economics. To link up an entire school system, cable must be strung above or buried below city streets, bypassing hundreds or thousands of homes. The taxpaying occupants of these homes get only an enhanced communications system benefiting the children — and possibly a tax reduction. Most adults will never use the system. A multipurpose system, on the other hand, is of value to the homeowner. By sharing the large capital costs, homeowners would aid cable viability, especially in urban areas. If the educational system were to make significant use of a cable system and pay a fair proportion of the costs, cable economics would be transformed. Instead of a heavy reliance on subscriber fees, a cable system would have an additional base of support — one larger perhaps than its commercial subscriber base.

With a new technological educational tool at its disposal, some school systems might reevaluate traditional spending priorities, opting for ITV hardware and software instead of bricks and mortar. It need not be an either-or choice, rather an appropriate mix of the old and the new. Technology can play a role, but it cannot do the entire job. ITV will permit a school system to use its resources more effectively, particularly its teaching staff. Class sizes may be reduced, allowing more time for individualized instruction. This has been the Anaheim and Hagerstown experience and there is no reason why it cannot be implemented elsewhere.

Obviously, one ITV teacher with a "class" of a thousand — or ten thousand — is an attractive economic proposition. The central issues then becomes one of evaluation. Which mode of instruction, conventional or ITV, produces the best results? The question is answerable by objective measurement; all that is needed is an open mind, a cable system, and a first-rate ITV curriculum.

Opposing educational theorists need only the courage of their convictions,

a willingness to take the results seriously, and a school system able and sufficiently interested to run the test. It is more than likely that foundation and/or
government money would underwrite this experiment, in whole or in part.
The most profound implications flow from the answer to the question: Can
students learn using ITV? If the answer is in the affirmative, the larger lobby
(the taxpayers) will prevail over the smaller lobby (the teacher's union). The
latter has been able to preserve the status quo only because no definitive
results have been adduced; into this vacuum, theory and precedent take root,
holding innovation at bay. Large as the issue of education is, it becomes part
of an even larger one: Will education be the catalyst of the communications
revolution? That could well turn out to be the case, particularly if the school
board of a city were to incorporate home study as part of its educational program. In this event, many options become available, some involving impressive opportunities for saving money, an authentic educational revolution. To employ this option, every home with school-age children would have
to be wired for CATV. With this communications capacity, the home as well
as the school could become a classroom — at least, for some courses. These
might be of the supplementary or enrichment variety, compensatory programs for students needing additional instruction, bilingual instruction for
preschool students (a significant problem in some school districts), optional
studies for gifted students, i.e., advanced science, mathemetics, computer
programming, etc. As an adjunct to conventional classroom instruction,
ITV via cable could provide many educational opportunities that would
otherwise be unavailable. The beneficiary of this system would be the student. But what about the long suffering taxpayer? Is this yet another costly
educational disaster? The answer is that it could be in the hands of the wrong
people. One should never underestimate the power of an entrenched and unsympathetic bureaucracy. On the other hand, an enlightened and committed
group of educators making full use of this communications system might
substantially upgrade the quality of education. Although the potential exists,
a communications system in itself guarantees nothing; the medium is not the
message. Only instructional programming of high quality is likely to produce
good results. A poor teacher is a poor teacher whether standing in front of a
class or in front of a camera. Only the best, most creative, and dedicated
teachers should be recruited for ITV.

As with all educational experiments, the cost-benefit equation merits close
examination. Would such programming pay for itself? To answer this question, several others must be aked: What is the scope and intent of the ITV
program? Is it to be a supplement or a substitute for conventional instruction? How many students are involved. What is the cost of the hardware?
The software? The operational costs?

In weighing the cost of ITV, which will be considerable, conventional education costs — current and projected — should be taken into account. With teacher costs escalating along with the capital costs of constructing schools, decision makers might take the process one step further: Why not opt for a home-study program? Why not use some of the time students have in the afternoon or evening to their advantage? Instead of playing video games or watching cartoons, why not learn history or science? The physical plant, i.e., the living room, playroom, or bedroom already exists; its use costs the taxpayer little or nothing in addition to what he is currently paying. Schoolrooms, on the other hand, are very expensive to build and maintain. Thus if students could be partially educated at home, the physical-plant requirements for the school system could be reduced. If, for example, half the students were being educated at home via ITV with the other half in school, a given facility could educate twice the number of students, in effect, double sessions via cable. Admittedly this idea will come as a culture shock to traditionalists who believe that physical confinement of students in a bricks-and-mortar fortress is the best, and perhaps the only way, to educate children. Without strong direction, control, and leadership, nothing of value, educationally speaking, is possible. This argument ignores the demonstrated capacity of television to teach when skillfully employed; it ignores the "education" that commercial television is giving to the young. With children spending more time watching television than they do in the classroom, does it not make sense to make better use of this time by teaching useful subjects? Would not many parents welcome this change in viewing habits?

Not all students could participate in a home-study program, since some come from homes where both parents work or from single-parent homes where there can be no supervision. These children would have to attend school, but even in these cases ITV could be employed to save money. Instead of a high-salaried teacher supervising the students, a teacher's aide could be used.

Once the schools and the homes of students in the district have been wired for cable, many scheduling options are possible: a week at home taking instruction, the next week in school; or two days at home and three days in school one week, three days at home and two days in school the following week. Other arrangements could be devised, including shorter school hours followed by televised instruction in the afternoon, the evening, or on Saturdays. Each school system could devise what it considered to be the proper educational mix consistent with its resources and educational philosophy. Obviously the school would continue to be the focal point in the educational progress; teachers would still teach, students would still learn; the objective would not and should not be compromised: the best education at the lowest cost. The debate — and most assuredly there would be a debate — will center on the question, How much television instruction and in what subjects? This is an appropriate question; television is not the answer to all our educational shortcomings. It should be used where it can be used effectively, where educational objectives are accomplished as well as or better than conventional instruction, and where savings can be made. Experience may dictate that some subjects do not lend themselves to the electronic media. Judgments ought to be made in this area by hard evidence, not theoretical prejudices. Televised instruction should be permitted to succeed or fail on its merit, predicated on careful research and objective data. Unsympathetic teachers and administrators should not be permitted to stack the deck. Evaluation of results and program quality ought to be subjected to independent review by experts outside the school system. Obviously, such a fundamental departure from educational precedent should undergo careful testing before being adopted by an entire school system. Education, a community's single most important and expensive obligation, is too important a matter to be left to chance. Despite the failure of our current educational system, innovative ideas must sustain the burden of proof; educators who want change had best do their homework.

With cities on the brink of bankruptcy, technology may yet have a hearing, an opportunity to prove what it can accomplish. For this to happen, some city will have to break new ground; some mayor or other elected leaders will have to have the courage to take on the educational community — administrators, teachers, and unions. The politics of change will be ferocious — of that, there can be little doubt. Clearly many teaching jobs will be eliminated, a development that will not be viewed with disinterest by the educational establishment. There will be strikes and threats of strikes, and predictions of educational disaster (larger even than the existing one).

Though teaching staffs will be markedly reduced along with a reduced physical plant — assuming a home-study plan is implemented — other costs will be incurred: cable hardware, software (ITV programming), technical staff to keep the system operational. It will not be an inexpensive operation; the start-up costs will be especially high. Once operational however, unit costs should decrease since film or videotaped programs are a *renewable* resource. Many programs in the curriculum are not subject to dating; once produced, they can be used for many years without cost to the school system. Other subjects are sufficiently topical so as to require periodic updating; even live programming might be employed. Judgments concerning the appropriate mode of instruction can be made on a case-by-case basis. Live programming has the advantage of spontaneity and immediacy, but it has the potential of being uneven in quality. Teachers like the rest of us have their good days and bad days. If, however, film or videotape is used, this factor is reduced or eliminated; the quest for the model lesson or model course of instruction is more easily attainable. With experience and objective test data, educators will be able to predict what programs produce results. Live programs are not so predictable, especially if there is a turnover in personnel; teachers who know how to use the medium as opposed to those who do not may be expected to be more effective. In opting to use only live programming, a school system restricts its selection of ITV personnel. The best teachers may be employed outside a school district, in another city or another state, at a local university or one hundreds of miles away. Live programming may make it impossible to secure the services of these teachers; the use of film or videotape may facilitate it. Obviously, no school system ought to settle for anything less than the very best education for its students, but if it elects or is pressured into an inbred, parochial approach, it may encourage mediocrity. It makes very little sense to buy the best technology and turn it over to inferior instructors.

The challenge of teaching thousands of students may prove irresistible to the superbly qualified teacher, a teacher's teacher, talent rarely available below the university level. With ITV, new educational opportunities are created. Local colleges and universities could use the cable system in a home-campus program. Private and parochial schools might use the system and share the costs. During the evening hours when most of the channels are unused, adult education courses could be offered. A metropolitan approach with resulting economies of scale might evolve. If ITV proves successful, satellite communities will not have to be persuaded to join. In education as in most endeavors, nothing succeeds like success.

Should the marriage of technology and education produce savings as its advocates expect, a community can elect to cut taxes or plow the money back into the schools in an effort to achieve excellence. Class sizes could be lowered, even one-to-one tutoring implemented, an approach that would win favor with teachers facing wholesale layoffs. With most communities struggling to keep from going under, this is the kind of decision they would welcome.

The National Education Association has been a strong advocate for the use of cable in education. In its booklet, *Schools and Cable Television,* it stated the following: "The NEA has consistently encouraged school districts and its local professional associations to explore the possibilities of using catv channels to extend the work of the schools by both in-school and out-of-school instructional programs via cable."[5]

Several colleges have used cable for instructional purposes. Oregon State University in Corvallis initiated fifty hours of ITV on campus with a feed to subscribers of cable in the towns of Corvallis, Albany, and Lebanon. Amarillo College in Amarillo, Texas, offered home-study courses via cable. In Bakersfield, California, two cable companies joined the Southern California Consortium in offering home-study courses.

If cable is to be used in home-study programs, every home with a school-age child — and perhaps preschool children, will have to be wired for cable. The third wire becomes part of the educational plant, not an optional service as it is for other subscribers. Similarly a television set must be provided, one for each student. While nearly every American home has at least one receiver, a home-study program cannot be predicated on its availability. The parents (and owners) may have other plans. Clearly there will be a need for a major investment in electronic hardware, and the cost would have to come out of the school budget. Though substantial, the cost of the basic cable service and the receiver should total less than two hundred dollars a year assuming a five-year life for a TV set costing four hundred dollars and seventy-five dollars for the cable service (one-way service). An interactive system with home terminals would be an additional cost.

Currently, test results on the relative merits of ITV vis-a-vis conventional instruction involves one-way television (i.e., instruction with the students not having the capacity to interact with the teacher). Traditionalists argue that this is a basic fault, test scores notwithstanding. Ideally, education should involve substantial interaction between teacher and student. If costs were not a factor, small classes of ten or a dozen might be expected to produce excellent results. Unfortunately, this kind of education is not possible in public education which must accommodate economic realities. Even when class sizes are large, some argue, student-teacher interaction exists if on a somewhat limited basis. One-way ITV eliminates this. There is certainly validity to this contention, but it would be incorrect to assert that *all* learning is inferior unless the student is in the physical presence of a teacher. Much of what we learn comes from a noninteractive experience; we learn by reading, by experiencing, by thinking — the education of living. One shudders to contemplate how profoundly ignorant we would all be if learning were confined to a structured academic environment.

At its best, an interactive educational experience can heighten a student's intellectual interests. An inspired teacher can make a difference in the lives of students; by force of example, he can lead the way for others, opening new doors and new possibilities. Sadly, there are too few of these teachers around these days; times have changed. Not only have the three R's suffered despite all the interactive teaching money could buy, but of even greater concern to parents and educators is the pervasive lack of interest in learning. Students have tuned out en masse. The joy of learning has become a joke — if a rather expensive one. Perhaps it is time for a new approach, an innovative departure from the present system. If the educational message is not getting through, perhaps a new medium will help.

As with any fundamental change — and the extensive use of ITV would be one — it can be taken to extremes thus defeating its purpose. Certainly, nobody would advocate an exclusive ITV program, one eliminating classroom instruction. Such an education might be a disaster, a generation of children reared on the tube without social interaction with their peers. What is the appropriate mix between classroom instruction and ITV? Could a student benefit from two or three hours of ITV at home or in school? (The average child watches more than five hours of commercial television daily.) Would a student *prefer* ITV to classroom instruction? If so, what effect if any, would this have on learning? What subjects lend themselves best to the electronic medium?

Educators at Hagerstown, Anaheim, and other systems have provided some but not all of the answers. We have much to learn. A general proposition however, may be stated: Any subject can employ ITV to advantage, even vocational education requiring manual skills. Becoming a good mechanic or carpenter involves more than turning a wrench or swinging a hammer. Often what needs to be learned can best be presented visually. Art lends itself to the electronic medium; one can travel to the great museums of the world via film or video tape. In the sciences one can look inside a single cell (biology) or see the universe (astronomy). Even abstract subjects in the hands of a master communicator can have vitality. An economics class taught by Milton Friedman or John Kenneth Galbraith would be absorbing — especially if each taught for a semester. Political science taught by Henry Kissinger would be all but irresistible.

ITV is not the answer to all our educational ills, rather part of the answer. We will need state-of-the-art hardware, a carefully constructed curriculum, and most important the very best teachers available. And it may be that the best ITV teachers are not professionally trained to teach; he or she may be a scientist, a business executive, a labor leader, a retired general, a mayor, a governor, a retired president of the United States. There are many people in our society willing and able to impart their particular knowledge; we should take advantage of this reservoir of talent. If this be educational heresy, as some may claim, we should make the most of it.

With few exceptions, ITV's accomplishments have resulted from the utilization of one-way systems. Proponents readily concede that an interactive system would be useful, even necessary in some courses of instruction; other courses require little or no student-teacher interaction. Monitoring a student's progress can be accomplished by means of assignments and periodic examinations. After years of experience, a teacher should have a rather good idea as to what areas of instruction cause difficulty. Students having problems could be assigned to watch an hour or two of additional instruction designed to explain the material in a new format and/or greater detail. If problems persist, the student could be assigned to a remedial class in school with intensive instruction, one with a low student-teacher ratio. Under the present system, specialized instruction is a luxury that many school systems cannot afford, the result being that the slow learner or those having special needs are permitted to fail.

Though one-way ITV has had good results, two-way ITV is clearly superior, providing a vastly expanded range of educational options. An interactive capacity would assist both the teacher and the student by closing the time gap between instruction and the awareness of a problem. If, for example, a teacher could know within seconds whether or not the subject had been fully understood by the student, he or she would be able to correct the matter immediately. It lends a degree of certainty to the educational process by insuring that the teacher never gets too far ahead of the students. A digital terminal linked to a central computer allows the teacher to give instant tests, true-false or multiple choice, and have a printout in seconds. If problems appear, the teacher can deal with them promptly. If the lesson is on film or video tape, a member of the teaching team can then appear live on the screen and offer additional instruction. (With a team-teaching concept, the ITV teacher and monitoring teacher need not be the same person.) When enough correct answers are given, the film or video tape can start again. If problems persist, a remedial lesson might be prepared with the uncomprehending students required to take it, perhaps as homework that evening.

The use of computers in an interactive cable system vastly expands ITVs usefulness. A teacher can know, literally on a minute-by-minute basis, how well the student has learned. The question that teachers constantly ask, Did you understand that? can be answered with precision. No longer can students hide their lack of comprehension for days, weeks, months. The ever-alert computer will not permit this. Students who fall behind, particularly in subjects where advanced instruction is predicated on early comprehension, not infrequently decide to give up. With an interactive system this cannot occur without the teacher and the school administration knowing about it. That most elusive concept, educational accountability, can become a reality. With testing on a lesson-by-lesson basis and a record of performance, teachers, administrators, and school boards will be under pressure to produce results. The computer printout is as much a test of professional competence as student comprehension.

The merciless computer will make demands on the student; inattention as well as a lack of comprehension will be readily evident. In many schools today students are merely obliged to be present physically; the mind can be anywhere. Every teacher has seen that glassy, far away look.

ITV can be a severe disciplinarian, demanding more of students than they might otherwise be prepared to give. As a mechanism for control, the computer has no equal. Students will try to beat it, of course, but they will lose every time. Each student will be assigned a number, a digital signature for each course. At the start of the lesson they will sign in and at the end sign out; in between they will be quizzed from time to time. If the student responds incorrectly to the true-false or multiple choice questions the computer will document this; if there is no response, the computer will detect it. Yes, but how will the computer know who is operating the terminal? It might be a friend or an older brother. True, but the stand-in would be obliged to answer every question and remain for the entire lesson. These kind of friends or relations are in short supply. Should such an attempt be made, the computer could be programmed to call attention to certain variables. If a student suddenly responded differently from his norm, a transformation from a C to an A student, that would be a red flag meriting inquiry. The student could be asked to take the test over — under supervision — to verify the result. From time to time, perhaps once or twice a month, all students would be tested under teacher supervision and the test scores compared. Eventually, the word will be passed that the system cannot be defeated and the realization sets in that one must study and learn, that there is no choice in the matter.

In an ITV program the physical setting in which a student learns is unimportant. It may be in a formal school environment (TV booth, earphones, and terminal) or at home in a favorite chair; one can learn at nine in the morning, at nine at night — or on a weekend. The system can have great flexibility for the student as well as the school system. Eventually, instead of an ITV schedule, some students may have access to an on-demand curriculum, any lesson, any course, at any time.

As with all options, ITV must be measured against conventional education both in effectiveness and cost-effectiveness. Clearly, there is need for additional study, for pilot programs and objective evaluation before any marriage between education and technology is arranged.

16

The Franchising Wars

The theory was to get the franchise, particularly in the large cities. Promise anything — large city fees, low subscriber rates, major sophisticated systems, two-way communications capability, (36 or more channels) — anything to get the franchise. Change it years later when the system was built and running. Use the franchise to raise new funds. It was standard procedure.
— Burt I. Harris, president of Harris Cable Corporation, 1975.

Wiring the major urban centers has always been the ultimate objective of the cable industry, an absolute prerequisite for it to become a major communications system. Rural communities and upscale suburbs provided a reasonable but unspectacular return on investment. Cable was a secure, minor-league business living in the shadow of its big brother, broadcast television, and destined to remain so until it entered the cities to compete. With only a few exceptions — San Diego and Manhattan — cable stayed out of the urban market. One reason was cost, more than $100,000 a mile in some parts of a city. A second more important consideration was the lack of demand. Most people were content with broadcast television, and cable had little alternative programming to sell. Given these conditions, cable operators prudently decided to wait. Most had no choice in the matter. Banks were not lending money to such speculative ventures, and cities were not granting franchises. Urban cable was an idea whose time had not yet come.

It was the emergence of Home Box Office as a major program supplier and the building of the QUBE system in Columbus that began to change the public perception about cable. Some of the potential of the medium became a reality. People took a closer look and liked what they saw. Cable could deliver on its promises, do things that broadcast television could not or would not do. The urban markets became receptive, indeed anxious, for cable. It was a remarkable turnabout. Bankers extended hundreds of millions of dollars in credit mostly to the large, well-established multiple-systems operators, those companies with the size and resources to bid for urban franchises. Small and medium-sized cable companies concentrated on the rural and suburban markets.

Mayors soon became bullish on cable — yet another conversion. In the early seventies most mayors had been as cautious as bankers about embracing cable; it was simply too risky. If they let cable in and it failed to attract sufficient patronage, rates would go up, and the mayors would be blamed. Like telephone and electric services, cable is a de facto cost-plus business that will not be allowed to go bankrupt; costs are simply passed along. Allowing cable in was to assume a political risk, one most mayors decided not to take.

Mayors were also mindful of cable operators' reputation for making promises to obtain a franchise and then later breaking those promises when it suited their purpose, knowing full well the leverage they enjoyed. Revoking the franchise for nonperformance was a costly, drawn-out remedy seldom employed; such matters were negotiated, often to the satisfaction of the operator.

Learning from the past, most cities took care in drafting their franchise agreements, employing expert legal counsel and knowledgeable cable advisers. In addition, most cities hired a professional staff and director to protect the city's interests on a continuing basis, overseeing the entire process from request for proposals to the operational stage.

Unlike broadcast-television licensing, cable franchising is a local prerogative. The FCC, once deeply involved in the process, has by court decisions and deregulation been largely eliminated as a force in cable, its powers markedly reduced. Except for broadcast-television interests, these developments have met with wide approval. The FCC, acting at the behest of the commercial broadcasting industry, was essentially hostile to cable, enacting a variety of rules — movie rules, sports rules, distant-signal rules, and so on — designed to protect broadcasters, keeping cable in a regulatory straight-jacket. These efforts slowed the development of cable but failed to stop it; the momentum was too powerful. Cable's time had come.

Because of local jurisdiction over cable, applicants for franchises could not work the levers of power in Washington to get what they wanted in the way that broadcast television interests had done in the late 1940s and 1950s. It was a new game with new rules and different players. There were, however, variations on an old theme with some innovative changes. Politically connected attorneys were engaged to represent cable applicants, and so-called rent-a-citizen tactics were devised whereby leading citizens bought stock or stock options in the cable corporation at a fraction of the real value — stock that could later be sold at a handsome profit if the applicant got the franchise. It was a sophisticated, deferred bribe in many instances. In return, the rented citizen was expected to use his or her influence. The pressure on cable commissions, city councils, and mayors was enormous; everyone who was anyone had a favorite cable company. The merits of a company's proposal often assumed secondary importance to its political clout. It was hardball, and everybody wanted to win.

With such high stakes, franchise applicants attempted to outpromise one another in a variety of areas: channel capacity (50 to 100, or more); two-way systems; access channels with staffs to train users, financed by dedicated funding; multiple pay-cable services; low basic rates, or even free universal service. The idea was to get the franchise; to make whatever promises were necessary.

Some observers are not entirely persuaded that those gaining franchises can or will honor all their commitments. There will be another phase according to Frank V. Greif, president of the National Association of Telecommunications Officers and Advisors. "The next wave could well be renegotiation — that is, when the cable company comes to the city and says, 'We agreed that after four years we'd have ten access studios and so on and so forth. We're only getting 40 percent penetration when we expected 60 percent. City, you're faced with a choice. Either we build the studios and go broke, or we back off and stay in business. What are you going to do?' I think it's real important for cities to start looking at that right now."[1]

Others entertain similar misgivings about paying for all those promises, one of them being Tony Hoffman, vice-president of A. G. Becker. With the high building costs of urban cable systems, Hoffman has predicted: "You're going to have to go back to the city and weasel out of everything you promised."[2]

Whether such circumstances come to pass remains to be seen. Much depends on urban cable's penetration rates both for basic and pay services. If it is in the 50 percent range for basic cable, companies should do well despite the large capital investments; if it is under 40 percent, there will be some distress. A few operators are getting nervous with all the money going out and little or nothing coming in. Others have lost neither their nerve nor their confidence. They expect a big payoff eventually and are untroubled by short-term growing pains. Most financial analysts and bankers remain cautiously optimistic as the cable industry moves forward into uncharted territory. The market for cable services is enormous, if unproven. Nobody can predict with certainty the level of acceptance for enhanced services. Do people really want to do their banking at home? Or shop at home? Subscribe to teletext/videotext services? Will they pay for fire-alarm and other security systems? There is a market for all these — and many other — enhanced services, but there is a wide range of opinion as to its potential especially in the near future. It may be an evolutionary rather than a revolutionary process.

For the moment, cable is very much an entertainment medium with pay cable as the moving force; it is making everything else possible. But for pay cable the question of enhanced services would be moot; there would be no enhanced services to speculate about. Indeed, there would be no urban cable revolution in progress. Although the national pay to basic ratio is 54.29 percent currently, that figure is for all cable, including rural obsolete systems that have no excess channel capacity to offer pay cable. In the urban and new-build systems an entirely different result is emerging. For every subscriber opting only for basic cable, two or more subscribe to pay services. While the national monthly average expenditure for cable subscribers is $17.17[3] (pay and basic), heavy users of pay cable can spend twice that amount or more every month. Subscribers of multiple-pay programming complain about duplication, with the same movie offered on two or three services. Feeling cheated, some disconnect what cable operators call churn. It is a major problem in many cable systems, one for which the operator is not responsible — nor is the program supplier for that matter. There is simply not enough quality programming to sell at this time with Hollywood making so few movies. The gap between demand and supply is enormous. To correct this deficiency, program suppliers like Home Box Office are underwriting movies for pay cable, a trend that seems likely to continue as the market for quality programs expands. Some predict a new golden age for Hollywood, something reminiscent of the way it was before the advent of television — a time when many people of all ages, not just young people, went to the movies. In the next few years when most cities have been wired for cable, there will be a theater in nearly every living room. The market will be so large that even the high production costs (ten million dollars for a modest movie to forty million or more for a blockbuster) will be manageable, particularly on a pay-per-view basis. Currently most movies on cable TV are offered on a block-purchase price, with the subscriber buying a monthly service, often not knowing what movies will be shown. Movie theaters do not sell tickets on this basis; their customers would not go along. Pay cable does it because in most systems there is no choice. The hardware needed for pay-per-view, address-able converters, has not been installed. The cable industry has had problems with the reliability of first-generation addressable converters, although Warner Amex systems work well. For the time being, most cable operators are prepared to wait until addressable converters are perfected.

Cable operators, program suppliers, and movie studios anticipate the time when a quality movie will recapture its investment and earn a profit in a single evening. Hollywood moguls view the prospects of pay-per-view with something approaching ecstasy, the best invention since film. By the end of the decade, projections indicate there will be forty million pay cable subscribers in the United States. If on a given evening 10 percent of these subscribers opt to see "Jaws 5" in the comfort of their living rooms for the bargain price of ten dollars, the gross will be forty million dollars. These number can generate a lot of creativity: "Jaws 6" is sure to follow.

Most major sporting events seem destined for cable either on a subscription or pay-per-view basis. Some teams have already made the move; others are waiting until the urban markets are wired. Like the Hollywood moguls, owners of major league sports teams expect a bonanza.

17

Cable Smorgasbord: Something For Everyone

You name the subject, there's probably a cable network devoted to it. And if there isn't — just wait a couple of days.

— CBS Development vice-president J. Roger Moody, 1982.

Cable television has kept its promise to deliver a wide range of diversified programming, doing what broadcast television can not or will not do. Broadcast television seeks the largest possible audience, and profits, largely ignoring specialized and less profitable programming. By so doing it neglects a significant minority of viewers who, given a choice, would watch something else.

Catering to the needs of specialized audiences is what cable television does best utilizing its many channels. This marketing concept is not new; specialized magazines have targeted small, special-interest readers for many years, some quite successfully, providing information that mass-circulation magazines do not print. Cable is an electronic magazine rack where people can pick and choose.

Obviously cable networks would like to attract as many viewers as possible, but they are prepared to accept a modest audience share. They must compete with one another and commercial broadcasters. The economics of cablecasting are quite different from broadcasting. Broadcasting draws large audiences, which produces large advertiser revenues, which allows for generous production budgets — advantages that unfortunately seldom translate into quality programming. Cablecasters have small audiences, relatively speaking; small production budgets: and limited advertising revenues for ad-supported services. (Currently there are thirty ad-supported cable networks and sixteen subscriber-supported networks.)

The viewing audience, necessarily fractionalized, can support only a limited number of cable networks. Indeed, the ad-supported networks present a gloomy picture. As of early 1983, only one — (Cable News Network) — is at or near the break-even point; all the others are operating at a loss.[1] So why, one may ask, do cable networks persist? The answer does not lie in the present outlook, but rather in the future potential for such cable networks. Collectively cable networks are gaining a larger audience share while the broadcast television networks are losing viewers at a steadily increasing rate. In 1976-1977, according to A.C. Nielsen, the three networks had a prime-time share of 91 percent. In 1980-1981, this had dropped to 83 percent.[2] With dozens of major cities being wired for cable, this trend is likely to continue. The advertising agency McCann Erickson projects a 76 percent three-network share in 1990.[3]

Despite a sea of red ink, cable-network executives believe that there is a bright future in an ever-expanding market. They are willing to suffer short-term losses for anticipated long-term gains. Clearly, some are destined to be disappointed. There is a winnowing process underway, and it has already claimed two notable casualties. The much admired CBS Cable, an advertiser-supported service, lasted a little more than a year, ceasing operations in November 1982 after sustaining heavy losses. It failed to attract sufficient advertising revenue, or a realistic prospect of getting adequate support, to sustain its high-quality cultural programming. A second victim, the subscriber-supported The Entertainment Channel, lasted only nine months and lost nearly fifty million dollars.[4] Like CBS Cable, The Entertainment Channel offered cultural programming. Though employing different funding mechanisms, each failed because it was unable to attract a sufficient number of viewers. Many observers are still persuaded that a market exists for cultural programming but that both CBS Cable and The Entertainment Channel entered prematurely, before cable penetration rates were sufficient to support this programming, particularly urban penetration rates. Had they waited a few years, one or both might well have survived.

Timing is all important in cable network services as the success of Home Box Office (HBO) and Ted Turner's efforts prove. Home Box Office began in November 1972, pioneering pay cable services to an untested market. For most of its life, HBO was a money losing business. Fortunately, it had the considerable resources of its parent, Time Incorporated, to sustain viability. HBO had a good idea and could afford to wait for cable to expand. As cable grew, so did HBO. It has become the dominant force in pay-cable programming with more than eleven million subscribers, seven million more than its closest competitor, Viacom International's Showtime. Getting in early and taking losses for several years proved to be the formula for ultimate success.

Ted Turner, another pioneer, made the same judgment about the market potential of cable. He was the first superstation cable programmer, the first to start an all-news cable service. Like HBO, Turner got in early and captured a large share of the market. WTBS has more than twenty-four million subscribers, more than any other service; CNN has more than sixteen million subscribers.[5]

The late arrivals like CBS Cable and The Entertainment Channel faced formidable competition. With many choices of cable-network offerings, subscribers had to be urged to spend their money on — or at the very least give their attention to — the particular service or else it would die. This selection process is ongoing, and there are certain to be more casualties.

The winners of this spirited competition are the cable subscribers, freed from the hegemony of broadcast television. Though the networks will not soon pass away, destined to serve a mass audience with lowest-common-denominator programs, one can turn the dial with a good opportunity of finding something worth watching. There is, or soon will be, a program designed to satisfy the most demanding or eclectic taste.

Cable news services have enjoyed wide acceptance. In addition to Cable News Network — the first entrant and present leader — there is a spin-off network, CNN Headline News and the Satellite News Channel. Reuters News View is also competing for a surprisingly large audience, more than twenty-eight million subscribers for the four networks as of December 1982.[6]

Sports networks are popular with cable subscribers led by the Entertainment and Sports Programming Network (ESPN) with more than nineteen million subscribers, followed by the USA Network with thirteen million.[7] In addition, there are six regional sports services: Action Sports Network, Bay City Interconnect, Madison Square Garden Network, PRISM/Philadelphia, PRISM/New England, and SportsChannel. To add to this sports coverage, there are the superstations with additional offerings: WTBS, WGN, and WOR.

Music has found a niche in cable with the Nashville Network, a joint venture between WSM Incorporated, owners of the Grand Ole Opry and Group W Satellite Communications. Rock music enthusiasts can tune in on Music Television (MTV) to see and to hear their favorite performers.

There is a Black Entertainment Television Network (BET), and for Spanish-speaking Americans, the Spanish International Network (SIN). Others can seek salvation and find it in the Eternal Word Television Network (EWTN), the National Christian Network (NCN), or National Jewish Television (NJT).

Secular subscribers who want to keep an eye on what Congress is doing for them or to them can watch Cable Satellite Public Affairs Network (C-SPAN), a service offering hour after hour of speeches from the podium of the House of Representatives. If one finds this experience depressing, cable offers relief: Turn to the Cable Health Network.

Cable doesn't offer all things to all people, not yet. But with all those hundred or more channel systems being built there may be a need for them eventually.

18

The Politics of Cable: Who Controls Whom?

Most subscribers think that cable television is a bargain, finding it qualitatively and quantitatively superior to broadcast television service. There is some grumbling about the cost and duplication in optional, subscriber-supported services, but by and large, customers are content. Some are absolutely delighted with the variety of programs available and the technical quality of the picture they receive. For many television has never been better.

Viewed strictly in entertainment terms, cable is a success. However, it is or soon will become more than an entertainment medium, and therein lies a problem obscured by all the glitter and glamour. History is repeating itself. The first electronic miracle, radio, came into being with very little public scrutiny or debate, an experience duplicated by broadcast television. Entertainment once again threatens to become an electronic tranquilizer relegating public policy considerations to a secondary position.

One hopeful development is a return to semantic reality. Unlike "free" television, cable services cost money. Nor are cable owners "public trustees"; they are entrepreneuers who have invested money to make money. The arrangements are more honest and less hypocritical. In return for the use of public and private property, subscribers get a new communications service if they are willing and able to pay for it. The agreement is incorporated in the franchise; everything is open and aboveboard. Or is it? What precisely are the people agreeing to and for how long? Who is to decide how and under what conditions this new system of communications is to be used and for whose benefit?

Many of these questions — and other similar ones — remain to be answered. There is more than a little ambiguity in the power relationships between the operator, the subscriber, and the community. Who controls whom is ultimately the central issue that is still to be resolved.

Obviously cable interests have some answers to these questions. Like any other businessman, a cable owner wants to make as much money as possible. To accomplish this objective he needs control, the right to make decisions, answerable to nobody except the stockholders. In most businesses this is the norm, but electronic mass communications is not just another business; there are substantial public interests involved. Resolving the conflicting private and public interests should be a matter of negotiation between the parties involved but that has not yet happened. Before the ink has dried on a franchise agreement, cable operators attempt to alter the terms fundamentally, not by negotiation with the community, but rather by sending their lobbyists to Washington for legislation that will eviscerate it. This tactical ploy, which worked well for broadcast television in the past, has now become the modus operandi for cable operators. When seeking a franchise, cable television executives project the image of enlightened, progressive, public-spirited businessmen. Once established in a community their objective is to stay. This is entirely understandable from an investment perspective, but the agreement calls for a fifteen-year relationship, a long enough time to recapture the capital investment and a profit. If all the promises are honored, the community may decide to renew the franchise for another fifteen years once it concludes that such an arrangement is in its interest. There is, however, no obligation.

While accepting this provision, cable operators are doing everything they can to subvert it, and they may well succeed. The National Cable Television Association (NCTA) is now a potent lobby, not yet in the same category as the National Association of Broadcasters (NAB) but rising rapidly. This is a reflection of the new realities in cable; it is now a force to be taken into account. A decade ago, cable television was an orphan. Nobody took it very seriously, especially not congressmen and senators beholden to commercial television's interests. Whatever victories cable won were in the marketplace and in the courts; the FCC and Congress, at the behest of the broadcasting industry, acted to restrain the growth of cable. Things have now changed, and today cable gets a respectful hearing. Indeed, the very broadcasting interests that years ago fought cable are now heavily involved as multiple-systems owners and cable programmers. Having failed to regulate or legislate cable into oblivion, they decided to buy in, believing it to be more virtuous — and more profitable — to be right rather than consistent. This new alliance has given cable a political clout it scarcely could have imagined a few years ago. Given this shift in power, communities and citizenry alike should keep careful watch over cable interests or risk another media coup. Unfortunately, the struggle for power between broadcast and cable interests, between community interests and cable interests, has received scant attention in the press and has also been largely ignored by broadcast television. When discussed at all, particularly on television, the true dimensions, depth, and public-policy consequences are rarely examined. It is a private contest waged behind the scenes, in congressional offices, subcommittee hearing rooms, and Washington cocktail parties. Most of us will only hear about it after a deal has been struck.

The latest legislative efforts to circumvent local jurisdiction over cable are worth examining inasmuch as they suggest the future. In 1980, S 2827 was introduced as an amendment to the Communications Act of 1934. Its provisions, supported by the National Cable Television Association, included the following: (1) local governments would be prohibited from regulating rates for basic cable service; (2) local governments could not require any program origination, including access programming for subscribers or a dedicated government channel. Both these provisions were to be retroactive. At the very time that cities and cable operators had reached agreement on access channels, studios, staff, even dedicated funding for public access, municipal channels, and basic cable rates, the industry's paid lobbyists and trade association representatives were attempting to render such agreements illegal.

In 1981, when the Senate was working on a rewrite of the Communications Act of 1934, S 898 was offered as an amendment to the act. It would have removed from local authority the right to regulate basic cable rates where "competition" existed (any community, in other words, that received broadcast television signals). Regulation was unnecessary, according to NCTA president Tom Wheeler, because it was "just one of many souces of entertainment and "not a necessary service."[1] The amendment passed the Senate by a vote of 90 to 4.

The board of directors of the National League of Cities (NLC) issued a resolution stating that it "strongly opposes federal legislation that would deny cities the right to regulate cable television through franchising agreements."[2]

New Orleans chief executive Ernest Morial called S 898 "damn poor public policy that would make me feel embarrassed if I were responsible for creating it." Cable operators would be free to do whatever they wanted , whenever they wanted to do it, for whomever they wanted to, and for however much they wanted to charge.[3]

In 1982, cable interests and their many friends in the Senate were again active. This latest effort — S 2172 — backed by Senator Barry Goldwater, addressed the sole issue that cable operators most fear — not having their franchise renewed. They want their fifteen-year relationship with a given community converted into a marriage, one from which divorce is vitually impossible. Cable interests seek what broadcasters have: a de facto right to operate forever in "their" market. Franchise renewals should be made automatic — a pro forma exercise as it is for television license renewal. The mechanism to accomplish this cherished objective is the "expectancy criteria" incorporated in S 2172. A renewal would be granted if the cable operator has substantially complied with the material terms of the franchise, if there has been no change in the legal, technical, or financial qualifications of the cable system that would impair service, and if the service meets the needs of the community and the system is state-of-the-art, compared to service being offered communities of similar size and characteristics.[4]

Under the Goldwater bill, a community would no longer be able to build its own cable system or enfranchise a nonprofit corporation to own and/or operate a cable system once it had granted a franchise to a private operator meeting the requirements of S2172. Once consummated, the marriage is forever — precisely what cable interests want. They have yet to achieve it. S2172 died in the Senate during the 1982 lame-duck session, but this phoenix is destined to rise again in 1983. Cable operators want it badly.

Were it not for the House of Representatives, S2172 and other amendments like it that are favorable to cable owners would now be the law of the land, inasmuch as a presidential veto seemed unlikely.

Leading the fight against the Goldwater bill, the National League of Cities (NLC) was joined by other public-interest groups. When details of S2172 became known, the NLC led a counterlobbying campaign. Cynthia Pols, legislative counsel for the NLC, was encouraged by the response: "We've never had an issue take off in terms of interest from our membership the way this one has, where everyone feels exactly the same way about the issue regardless of their party affiliation or ideological bent."[5]

The issue as seen from the NLC perspective is one of local control over cable, a matter many had assumed was settled. After several legal setbacks, the FCC largely abandoned its regulatory role in favor of local control, a move that won wide approval from cable operators and communities freed from regulatory constraints. At last the marketplace would be allowed to function without the help of Big Brother. Cable operators and communities could negotiate whatever arrangements they found mutually satisfactory.

With cable's back-door legislative efforts to get what they could not gain through negotiation, the relationship between the industry and local communities has deteriorated. Mayors and their communications advisers have taken a second look at their franchises, wondering which provisions may be rendered null and void by Congress. Many communities spent a year or more carefully researching, negotiating, and drafting a franchise agreement, detailing the rights and obligations of both parties. Now, it seems, they must be prepared to renegotiate. The center of power has shifted, once again, back to Washington.

To add to the confusion, the United States Supreme Court has yet to define the precise First Amendment rights of the cable operator vis-a-vis those of the subscriber and the local community. Does cable have the same rights as radio or television broadcasting? As the press? No one is quite certain. One case that has reached the Supreme Court, *Midwest Video Corporation* v. *FCC* (571 F.2nd 125; 440 U.S. 689) strongly suggests that cable operators do enjoy First Amendment rights, although the case was decided on narrow grounds. Midwest Video Corporation sued the FCC claiming that the access rules for subscribers violated its First Amendment rights. While not deciding the issue, the Supreme Court went out of its way to declare that a First Amendment challenge of the cable operator was "not frivolous."

Thus, it is conceivable that some day a frontal assault on access channel use may be made in the courts with the outcome uncertain. If the Supreme Court decides that cable operators and broadcasters stand on the same legal ground, access as a *matter of right* for subscribers will be a thing of the past, thousands of franchise agreements notwithstanding. Of course, cable operators like broadcasters may grant their subscribers access as a matter of policy, but it may be subject to censorship. Cablecasting could become like broadcasting, a chilling prospect for those who had such high hopes for cable, the first mass-communictions medium accessible to all, an authentic marketplace for ideas.

While the general public is enjoying the video feast, some are looking back over both shoulders to see who is gaining — the judges or the politicians. Belatedly, it has become evident that most communities have spent too much attention on the wrong issues: channel capacity, interactive technology, basic cable rates, franchise fees, public-access channels and studios — the nuts and bolts of cable franchising — and far too little on how to protect themselves and their citizenry from political-legal risks. It is a wait and see time now and it need not have been. The Midwest Video case decided by the Supreme Court in 1979 before most urban franchises were awarded, should have been the red flag on access provisions of the franchise. Nor should the political games that cable operators play have come as a surprise; they are not less adept than broadcasters in working the Washington levers of power. All the present threats were entirely predictable.

Not everyone was surprised or taken in by the cable companies. Those who had done their homework viewed cable operators and their promises with skepticism. Some advocated municipal ownership, nonprofit ownership, and options to purchase as mechanisms that would afford a measure of protection from private interests. In recent years some eighty cities have explored municipal ownership. It is not a new idea. Frankfurt, Kentucky, opted for municipal ownership in 1952 and currently has more than eighty-one hundred subscribers.[6]

Other communities have chosen the cooperative concept, each subscriber being a shareholder in the cable system. There are at least sixty subscriber-owned systems operating, the largest being in Harlan, Kentucky.[7] Most of these systems are located in rural areas, many with fewer than a hundred subscribers. In Davis, California, however, the City Council has awarded a conditional cable franchise to the Davis Cable Cooperative, Inc., to build an urban cable system. The people in Davis who use the cable system will own it and set policy.[8]

There are some problems with municipal ownership. People are concerned

— and properly so — by the prospect of unwittingly permitting politicians to gain control of mass communications. It is a risk but no less real than that of political control by private interests, our current system of broadcasting that has acquired unmerited legitimacy. A board of directors chosen by subscribers can be turned out of power at the next election, something that cannot be done to a cable company's board.

Another option is for the community to own the system's hardware, turning management functions over to professionals on a contract basis with policy decisions made by an elected advisory board, all of whom would be subscribers. Should the political problems of municipal ownership appear too formidable, the community could award the franchise to a nonprofit corporation with its board of directors elected by the subscribers. The corporation could hold title to the physical assets and hire a cable management firm to run the system. With low-cost money — municipal bonds — and no need to make a profit — a minimum of 15 percent — such a system would have some obvious economic advantages over a privately owned system.

Should these alternatives be rejected, a community can still take measures to protect its interests in dealing with cable operators. One is an option to purchase or, under certain agreed conditions, to renegotiate the franchise. For example, a community could exercise its option if a substantive provision of the franchise is rendered null and void by legislative act or judicial decision. If every franchise had this escape clause, cable operators might entertain second thoughts about lobbying or litigating. The franchise would truly represent the intent of the parties, not merely a temporary tactical exercise.

To ask who should own the hardware is to pose a second question: Who should have the power? In the United States of America, property is power, and particularly so in the field of mass communications. If cable is not to become another instrument of private power, if its promise is to be fully realized we must be alert to the dangers.

19

Cable and the First Amendment

"It's our investment and our channel," Marty Lafferty, Cox Cable's director of programming services, explained to a group of newspaper publishers who were inquiring about their right to use cable. His answer was simple: They had no rights. Neither had any right of access to the other's medium. Cox Cable, Lafferty went on to say, reserved the right to preview any material disseminated by its cable system.[1]

The newspaper publishers were incensed. "Suppose cable becomes the principal way of delivering newspapers. What will happen to freedom of expression?" asked Ralph Lowenstein, dean of the University of Florida's Journalism and Communications School.

Lafferty was unmoved. Cox had an obligation to shield its subscribers from "objectionable content," he blandly declared.

The publishers were understandly outraged. Cox was asserting the right of the print media to exclude whatever or whomever it chose — an intolerable situation. What arrogance!

This confrontation between the old and new movers and shakers of the communications universe would be amusing were it not so serious. It points up the issue of free access to cable by information suppliers and others wanting to sell software. If the Cox view prevails — the gatekeeper approach — subscribers will see only what Cox allows them to see. Sellers of services will pay what Cox demands or go elsewhere; subscribers will pay what Cox demands or go without the service. So whatever happened to the free market of ideas that cable operators originally promised?

The real issue is, What should be the legal status of a cable system? Should it enjoy the rights of a monopoly newspaper? Of broadcast television station? Or perhaps something quite different given its special character?

If you want to get a cable operator exercised, say the magic words *public utility.* Despite overwhelming evidence to the contrary, the industry contends it is not and should not be regulated as a public utility. It offers a desirable but not a necessary service; there is competition: broadcast television, subscription television in some markets, and other delivery systems that will eventually come on line; direct broadcast satellites, satellite master antenna television, and multipoint distribution service.

Predicated exclusively as an entertainment medium, the argument would not be without merit, but cable is not just an entertainment medium, even though it is widely promoted on that basis. This is an industry smoke screen designed to obscure the larger reality — and thus far the disguise seems to have worked.

Proponents of public-utility classification argue that cable is a de facto monopoly, nonexclusivity clauses in cable franchises notwithstanding. In such a monopoly environment, costs for enhanced services — nonentertainment — would rise artificially without the moderating affect of competition. More important, there are profound consequences in allowing the medium to control the message.[2]

If cable had the status of a public utility, it would be obliged to lease channels on a nondiscriminatory basis, exercising no control over content — the role that the telephone company plays. The result of such a separation of functions would be a free marketplace both for entertainment and nonentertainment services. Cable companies would make money by renting or leasing their facilities.

The cable industry is opposed to playing so limited a role. It wants, and has succeeded in getting, almost total control. Aside from concessions negotiated with the community (access channels for public use, municipal channels, basic cable rates, etc., etc.) the cable operator can exclude whomever or whatever he chooses and charge anything he wants. As gatekeeper, he alone decides what we are permitted to see. For the time being, these decisions seem to have no more sinister motive than profit maximization, but who can say when ideology will be the criterion?

Several of the larger multiple-systems operators are also pay-program suppliers: Warner Amex, Cox, Times Mirror, Viacom, owner of Showtime, the second largest pay-cable service, and Time Inc., owner of Home Box Office, the largest pay-cable service, and ATC the largest multiple systems operator.

Understandably, MSOs who are program suppliers have more than a passing interest in what pay programs appear on its system — a replication of the vertical integration once part of the movie industry when theaters were owned by movie studios.[3] This distinctly unhealthy arrangement was struck down by the courts. Cable-system owners who are also program suppliers may yet have their day in court, but for the moment anything goes.

The result of such vertical ownership was highly predictable. Why should one MSO market a competitor's pay-programming service at the expense of his own? It made no economic sense unless there was money to be made, more from a competitor's service than from his own pay service. Some operators have resisted the temptation to preempt the market, preferring to allow many services and keep subscribers happy, thereby avoiding a rather delicate issue. Others take a harder line. It is our investment and our channel, and we will use it as we please, and it does not please us to allow our captive audience to enrich the competition. If they want pay programs we will provide an in-house service. The free marketeers of cable do not want the market to be *that* free.

When Times Mirror, a large MSO, initiated its Spotlight pay-cable service in May 1981, it decided to drop Home Box Office and Showtime, ranked one and two in pay cable. Many of the 170,000 HBO and 60,000 Showtime subscribers protested. Times Mirror held out for six months then capitulated, reinstating HBO in the twenty-seven systems from which it had been removed.[4]

In August 1982, Cox Cable announced it was dropping The Movie Channel, Cinemax, and Home Box Office from twenty-six of its systems, replacing them with Spotlight. Cox, Tele-Communications Incorporated, and Storer Communications had become joint owners of Spotlight with Times Mirror. Cox announced it was a decision intended "to provide their subscribers with the highest quality and best value in home entertainment."[5]

These and other market manipulations by cable systems may seem unimportant at first glance, until one considers the implications. The power to decide what movies are exhibited is the power to restrict *all* programs on system-controlled channels. Despite the huge capacity of cable, some programs will not be shown even though channels are available and exhibitors are willing to pay for them. The gatekeeper will not allow it. This is hardly the free marketplace of ideas, many had hoped for and expected.

The struggle for control of the mass communications system of the future is on-going. It is a struggle many Americans are unaware of. On its outcome much depends.

□ □ □

Notes

Chapter 1

1. Harry J. Skornia and Jack Kitson, ed., *Problems and Controversies in Television,* (Palo Alto, Calif.: Pacific Books, 1968), pp. 53-58.

Chapter 2

1. Herbert Hoover, *Memoirs: The Cabinet and the Presidency,* Vol. 2, 1920-1933 (New York: Macmillan, 1952), p. 140.
2. Bernard Schwartz, *The Professor and the Commission,* (New York: Alfred A. Knopf, 1959), p. 151.
3. Ibid., p. 151.
4. James M. Landis, *Report on Regulatory Agencies to the President-elect* (Washington, D.C.: Government Printing Office).
5. Stan Opotowsky, *TV: The Big Picture* (New York: E.P. Dutton, 1961), p. 102.

Chapter 3

1. Jerome A. Barron, *Freedom of the Press For Whom?* (Bloomington, Indiana University Press, 1973).

Chapter 4

1. Congressional Research Service of the Library of Congress, *Congress and Mass Communications* (Washington, D.C., Government Printing Office, 1974), p. v.
2. Op. Cit., p. 20.
3. Op. Cit., p. 20.

Chapter 5

1. Harry J. Skornia, *Television and Society* (New York: McGraw-Hill, 1965), p. 99.

Chapter 7

1. Fred W. Friendly, *Due to Circumstances Beyond Our Control* (New York: Random House, 1967), p. 3.

2. Ibid., p. 3.
3. Ibid., p. 59.
4. Ibid., p. I.

Chapter 8

1. *In the Matter of Television Station WCBS-TV, New York, New York* (Application of the Fairness Doctrine to Cigarette Smoking), 9 FCC 2nd 921 (1967).
2. *Friends of the Earth and Garie A. Soucie* v. *Federal Communications Commission of the United States of America,* 449 F2nd 1164 (1971).
3. *Fairness Report,* 48 FCC 2nd 1, 26 (1974).

Chapter 10

1. Marvin Barrett, ed. *Moments of Truth?* (New York: Fifth Alfred I. duPont-Columbia University Survey of Broadcast Journalism, Thomas Y. Crowell Company, 1975), p. 130.
2. Ibid.
3. Report of the Carnegie Commission on the future of public broadcasting, *A Public Trust,* (New York: Bantam Books, 1979), p. 49.
4. Ibid., p. 141.
5. Ibid.
6. Ibid., p. 140.

Chapter 13

1. Cablevision, Plus, August 2, 1982, p. 4.
2. *Cable Television Report and Order,* Docket 18397, p. 120 (1972).
3. *TVC,* December 1981, p. 70.
4. *Television Factbook,* 1979, services vol., p. 83a.
5. *Cablevision,* October 4, 1982, p. 112.
6. Ibid.
7. *Cablevision,* September 13, 1982, p. 195.
8. Ralph Lee Smith, *The Wired Nation,* (New York, Harper & Row, 1972).
9. *Cablevision,* September 13, 1982, p. 195.
10. Ibid., p. 195.

Chapter 15

1. Washington County Closed Circuit Television Report, undated, Board of Education, Washington County, Hagerstown, Maryland.
2. Cost Study of Educational Media Systems and Their Equipment Components, General Learning Corporation, 1968.
3. What We Know About Learning From Instructional Television, page 53. Wilbur Schramm. Stanford: The Institute for Communications Research, 1962.
4. Cable Television: Developing Community Services, page 139. Polly Carpenter-Huffman, Richard C. Kletter and Robert K. Yin. The Rand Corporation, Crane, Russak & Company, N.Y. N.Y. 1975.
5. Schools and Cable Television, National Education Association, p. 5, booklet, 1971.

Chapter 16

1. *Cablevision,* November 22, 1982, p. 239.
2. Ibid., p. 244.
3. Op. Cit., October 4, 1982.
4. Op. Cit., January 17, 1983, p. 77.

Chapter 17

1. *Time,* March 7, 1983, p. 69.
2. *Cablevision,* January 25, 1982, p. 84.
3. Ibid., p. 84.
4. *Time,* March 7, 1983, p. 69.
5. *Cablevision,* December 27, 1982, p. 104.
6. Ibid.
7. Ibid.

Chapter 18

1. *TVC,* August 15, 1981, p. 21.
2. Ibid., p. 20.
3. Ibid.
4. *Cablevision,* October 11, 1982, p. 65.
5. Ibid., October 4, 1982, p. 86.
6. *Television Factbook,* 1981-1982, services vol., p. 811a.
7. Op. cit., 1979, cable section, services vol.
8. *Channels,* March-April 1983, p. 14.

Chapter 19

1. *Cablevision,* September 13, 1982, p. 62.
2. Cabinet Committee on Cable Communications, *Cable Report to the President,* January 14, 1974.
3. *U.S.* v. *Paramount Pictures,* 334 U.S. 131 (1948).
4. *Cablevision,* November 2, 1981, p. 19.
5. Ibid., August 30, 1982, p. 18.

Selected Bibliography

Baer, Walter S. *Cable Television: A Hand-book For Decision-Making.* Santa Monica: Rand Corporation.

Barnouw, Erik. *A History of Broadcasting in the United States.* Vol. 1, *A Tower in Babel;* Vol. 2, *The Golden Web;* Vol. 3, *The Image Empire.* New York: Oxford University Press, 1966-1970.

_____ *Tube of Plenty.* New York: Oxford University Press, 1975.

_____ *The Sponsor.* New York: Oxford University Press, 1978.

Barrett, Marvin, ed. *Moments of Truth?* New York: Thomas Y. Crowell Company, 1975.

Barron, Jerome A. *Freedom of the Press For Whom?* Bloomington: Indiana University Press, 1973.

Bernstein, Marver H. *Regulating Business by Independent Commission.* Princeton: Princeton University Press, 1955.

Carnegie Commission on the Future of Public Broadcasting. *A Public Trust.* New York: Bantam Books, 1979.

Carpenter-Huffman, Polly, Kletter, Richard C. and Yin, Robert K. *Cable Television: Developing Community Services.* The Rand Corporation. Report done for Rand Corp. then published in book form by Crane, Russak. New York: Crane, Russak & Company, 1975.

Cole, Barry, and Oettinger, Mal. *Reluctant Regulators.* Reading, Mass.: Addison-Wesley, 1978.

Congressional Research Service. *Congress and Mass Communications,* Washington, D.C. Government Printing Office, 1974.

Connochie, T.D. *TV in Education and Industry.* Vancouver: Mitchell Press Ltd., 1969.

Devol, Kenneth S., ed. *Mass Media and The Supreme Court,* 2nd ed. New York: Hastings House, 1976.

Diamond, Edwin. *The Tin Kazoo: Television, Politics & the News.* Cambridge, MIT Press, 1976.

Diamond, Robert M., ed. *A Guide to Instructional Television,* New York: McGraw-Hill Book Company, 1964.

Epstein, Edward Jay. *News From Nowhere* New York: Random House, 1973.

Erickson, Carlton W. *Fundamentals of Teaching With Audiovisual Technology.* New York: Macmillan, 1972.

Friendly, Fred W. *Due To Circumstances Beyond Our Control.* New York: Random House, 1967.

_____ *The Good Guys, The Bad Guys and The First Amendment.* New York: Random House, 1975.

Geller, Henry. *The Fairness Doctrine in Broadcasting.* Santa Monica: Rand Corporation, 1973.

Goldsen, Rose K. *The Show and Tell Machine.* New York: Dial Press, 1975.

Head, Sidney W. *Broadcasting in America.* Boston: Houghton Mifflin Company, 1972.

Hollowell, Mary L., ed. *Cable/Broadband Communications Book,* Vol. 2. Washington, D.C.: Communications Press, 1981.

Johnson, Leland L., and Botein, Michael. *Cable Television: The Process of Franchising.* Santa Monica: Rand Corporation, 1973.

Johnson, Nicholas. *How To Talk Back to Your Television Set.* Boston: Atlantic-Little, Brown, 1969.

Klein, Paul, et al. *Inside the TV Business.* New York: Sterling Publishing, 1979.

Kohlmeier, Louis M. *The Regulators: Watchdog Agencies and the Public Interest.* New York: Harper & Row, 1969.

Levin, Harry J. *Broadcast Regulation and Joint Ownership.* New York: New York University Press, 1960.

MacNeil, Robert. *The People Machine: The Influence of Television on American Politics.* New York: Harper & Row, 1968.

Mankiewicz, Frank, and Swerdlow, Joel. *Remote Control: TV and the Manipulation of American Life.* New York: Times Books, 1978.

Martin, James. *The Wired Society.* Englewood Cliffs, N.J.: Prentice-Hall, 1979.

Mayer, Martin. *About Television.* New York: Harper & Row, 1972.

Mayer, Mary P. *CATV: A History of Community Antenna Television.* Evanston, Ill.: Northwestern University Press, 1972.

Melody, William. *Children's TV; the Economics of Exploitation.* New Haven: Yale University Press, 1973.

Michelson, Sig. *The Electronic Mirror: Politics in an Age of Television.* New York: Dodd, Mead, 1972.

Miller, Marie Winn. *The Plug-in Drug.* New York: Viking Press, 1977.

Minow, Newton N. *Equal Time: The Private Broadcaster and the Public Interest.* New York: Atheneum, 1964.

Opotowsky, Stan. *TV: The Big Picture.* New York: E.P. Dutton, 1961.

Paley, William S. *As It Happened: A Memoir.* Garden City, N.Y.: Doubleday, 1979.

Pilnick, Carl, and Baer, Walter S. *Cable Television: A Guide to the Technology.* Rand Corporation, 1973.

Pool, Ithiel de Sola. *Talking Back: Citizen Feedback and Cable Technology.* Cambridge: MIT Press, 1973.

Schiller, Herbert. *Mass Communications and American Empire.* New York: Augustus Kelley, 1969.

Schwartz, Bernard. *The Professor and the Commissions.* New York: Alfred A. Knopf, 1959.

Shapiro, Andrew O. *Media Access.* Boston, Little, Brown, 1976.

Skornia, Harry J. *Television and Society.* New York: McGraw-Hill, 1965.

──────── and Kitson, Jack William. *Problems and Controversies in Television and Radio.* Palo Alto: Pacific Books, 1968.

Sloan Commission on Cable Communications. *On the Cable: The Television of Abundance.* New York: McGraw-Hill, 1971.

Smith, Ralph Lee. *The Wired Nation.* New York: Harper & Row, 1972.

Television & Cable Factbook, 1979-1983 eds. Washington, D.C.: Television Digest.

Wolf, Frank, *Television Programming For News and Public Affairs.* New York: Praeger, 1972.

Woodward, Charles C., Jr. *Cable Television Acquisition and Operation of CATV Systems.* New York: McGraw-Hill, 1974.

Periodicals:
>   *Cablevision*
>   *Channels*
>   *TVC*
>   *Broadcasting*

Major Market Cable Survey
City of: Phoenix, Arizona

1. To whom was the cable franchise(s) awarded? *American Cable Television, Camelback Cablevision and Western Cablevision.*
2. For how many years? *15 years each.*
3. Total number of channels? *54 each.*
4. Does system have an interactive capacity? *Yes.*
5. How many tiers offered? *Two basic tiers each.*
6. What is the projected cost of the system? *$100 million plus.*
7. How many homes will system pass? *300,000.*
8. What is the cost of basic cable service? *$9.45, $8.00, $8.50.*
9. How many channels offered in basic cable service? *25 to 27.*
10. What is the municipal fee? *5% of gross.*
11. Percent of construction completed? *80%.*
12. Is construction on, ahead, or behind schedule? *Ahead.*
13. Date construction is scheduled for completion? *Dec. '84.*
14. Number of public access channels? *1.*
15. Is there a dedicated fund for access programming? *No.*
16. Will candidates for public office be permitted to use public access channels as a matter of right? *Undecided.*
17. Have channels been reserved for municipal use? If so, how many? *1.*
18. How many channels have been reserved for educational use? *1.*
19. Did your community consider awarding the franchise(s) to a non-profit corporation? *No.* Did it consider municipal ownership? *No.*
20. Is there a provision in your franchise(s) for leased-channel access by commercial program suppliers? *Yes.*

Major Market Cable Survey
City of: San Diego, California

1. To whom was the cable franchise(s) awarded? *(A) Cox Cable; (B) Time, Inc.*
2. For how many years? *Cox 20 years; Time 15 years.*
3. Total number of channels? *35.*
4. Does system have an interactive capacity? *Yes, experimental.*
5. How many tiers offered? *One.*
6. What is the projected cost of the system? *?*
7. How many homes will system pass? *?*
8. What is the cost of basic cable service? *Cox $11.25 a month; Time $11.39 a month.*
9. How many channels offered in basic cable service? *20 plus.*
10. What is the municipal fee? *3% of gross.*
11. Percent of construction completed? *Virtually all required except new construction.*
12. Is construction on, ahead, or behind schedule? *On.*
13. Date construction is scheduled for completion? *New construction ongoing.*
14. Number of public access channels? *1.*
15. Is there a dedicated fund for access programming? *No.*
16. Will candidates for public office be permitted to use public access channels as a matter of right? *Yes.*
17. Have channels been reserved for municipal use? If so, how many? *1.*
18. How many channels have been reserved for educational use? *1.*
19. Did your community consider awarding the franchise(s) to a non-profit corporation? *No.* Did it consider municipal ownership? *No.*
20. Is there a provision in your franchise(s) for leased-channel access by commercial program suppliers? *No.*

Major Market Cable Survey
City of: Denver, Colorado

1. To whom was the cable franchise(s)*
   awarded? *ATC and Daniels and
   Associates.*
2. For how many years? *15.*
3. Total number of channels? *60 plus
   separate Institutional Network.*
4. Does system have an interactive capacity? *Yes.*
5. How many tiers offered? *4.*
6. What is the projected cost of the
   system? *$78-$100 million.*
7. How many homes will system pass?
   224,000.
8. What is the cost of basic cable service?
   $2.50 per month.
9. How many channels offered in basic
   cable service? *33.*
10. What is the municipal fee? *5% gross
    revenues.*
11. Percent of construction completed?
    Less than 5% as of March 31, 1983.
12. Is construction on, ahead, or behind
    schedule? *On.*
13. Date construction is scheduled for
    completion? *August 1986.*
14. Number of public access channels? *12.*
15. Is there a dedicated fund for access
    programming? *Yes.*
16. Will candidates for public office be
    permitted to use public access channels
    as a matter of right? *No.*
17. Have channels been reserved for
    municipal use? If so, how many? *1.*
18. How many channels have been re-
    served for educational use? *3.*
19. Did your community consider award-
    ing the franchise(s) to a non-profit cor-
    poration? *No.* Did it consider
    municipal ownership? *Yes.*
20. Is there a provision in your franchise(s)
    for leased-channel access by commer-
    cial program suppliers? *Yes, at least 1
    channel.*

*In Denver a *Permit* not a franchise is
granted.

Major Market Cable Survey
City of: St. Petersburg

1. To whom was the cable franchise(s)
   awarded? *1971, TM Communications;
   1972, sold to Teleprompter; April
   1982, sold to Group W.*
2. For how many years? *20.*
3. Total number of channels? *Originally
   20, now 35.*
4. Does system have an interactive capacity? *Not yet.*
5. How many tiers offered? *2.*
6. What is the projected cost of the
   system? *N.A.*
7. How many homes will system pass?
   90,000.
8. What is the cost of basic cable service?
   $7.95/mo.
9. How many channels offered in basic
   cable service? *12.*
10. What is the municipal fee? *10%.*
11. Percent of construction completed?
    Complete.
12. Is construction on, ahead, or behind
    schedule? *N.A.*
13. Date construction is scheduled for
    completion? *Completed 1976.*
14. Number of public access channels? *1.*
15. Is there a dedicated fund for access
    programming? *No.*
16. Will candidates for public office be
    permitted to use public access channels
    as a matter of right? *No.*
17. Have channels been reserved for
    municipal use? If so, how many? *10%.*
18. How many channels have been re-
    served for educational use? *2.*
19. Did your community consider award-
    ing the franchise(s) to a non-profit cor-
    poration? *No.* Did it consider
    municipal ownership? *No.*
20. Is there a provision in your franchise(s)
    for leased-channel access by commer-
    cial program suppliers? *No.*

Major Market Cable Survey
City of: Tampa, Florida

1. To whom was the cable franchise(s) awarded? *Tampa Cable Television (A Tribune Company).*
2. For how many years? *15.*
3. Total number of channels? *122.*
4. Does system have an interactive capacity? *Yes.*
5. How many tiers offered? *3.*
6. What is the projected cost of the system? *54-84 million dollars.*
7. How many homes will system pass? *115,000.*
8. What is the cost of basic cable service? *$2.25.*
9. How many channels offered in basic cable service? —
10. What is the municipal fee? *3%.*
11. Percent of construction completed? *0 — construction begins fall 1983.*
12. Is construction on, ahead, or behind schedule? —
13. Date construction is scheduled for completion? *6/87.*
14. Number of public access channels? *17 allocated for public, educational, government and leased-access.*
15. Is there a dedicated fund for access programming? *Yes.*
16. Will candidates for public office be permitted to use public access channels as a matter of right? *Absolutely not!*
17. Have channels been reserved for municipal use? If so, how many? *Yes, see #14.*
18. How many channels have been reserved for educational use? *2.*
19. Did your community consider awarding the franchise(s) to a non-profit corporation? *No.* Did it consider municipal ownership? *Yes.*
20. Is there a provision in your franchise(s) for leased-channel access by commercial program suppliers? *Yes.*

Major Market Cable Survey
City of: Orlando, Florida

1. To whom was the cable franchise(s) awarded? *ATC.*
2. For how many years? *20.*
3. Total number of channels? *30.*
4. Does system have an interactive capacity? *Experimental.*
5. How many tiers offered? *2.*
6. What is the projected cost of the system? *Unknown.*
7. How many homes will system pass? *190,000.*
8. What is the cost of basic cable service? *$8.00.*
9. How many channels offered in basic cable service? *13.*
10. What is the municipal fee? *Minimum of $125,000 @ year.*
11. Percent of construction completed? *92.*
12. Is construction on, ahead, or behind schedule? *On.*
13. Date construction is scheduled for completion? *N/A.*
14. Number of public access channels? *1.*
15. Is there a dedicated fund for access programming? *Yes.*
16. Will candidates for public office be permitted to use public access channels as a matter of right? *Yes.*
17. Have channels been reserved for municipal use? If so, how many? *2.*
18. How many channels have been reserved for educational use? *1.*
19. Did your community consider awarding the franchise(s) to a non-profit corporation? *No.* Did it consider municipal ownership? *No.*
20. Is there a provision in your franchise(s) for leased-channel access by commercial program suppliers? *Yes.*

Major Market Cable Survey
City of: Indianapolis

1. To whom was the cable franchise(s)*
   awarded? *Indpls Cable & ATC.*
2. For how many years? *15.*
3. Total number of channels? *Indpls 27;
   ATC 37.*
4. Does system have an interactive capacity? *ATC — Yes.*
5. How many tiers offered? *Four.*
6. What is the projected cost of the
   system? *Indpls 9M; ATC 11M.*
7. How many homes will system pass?
   280,000.
8. What is the cost of basic cable service?
   Indpls 9.95; ATC 7.95.
9. How many channels offered in basic
   cable service? *Indpls 2; ATC 33.*
10. What is the municipal fee? *3%.*
11. Percent of construction completed? *In-
    dpls 100%; ATC 55%.*
12. Is construction on, ahead, or behind
    schedule? *On.*
13. Date construction is scheduled for
    completion? *2-19-84.*
14. Number of public access channels? *In-
    dpls 1; ATC 3.*
15. Is there a dedicated fund for access
    programming? *No.*
16. Will candidates for public office be
    permitted to use public access channels
    as a matter of right? *?*
17. Have channels been reserved for
    municipal use? If so, how many? *One.*
18. How many channels have been re-
    served for educational use? *Four.*
19. Did your community consider award-
    ing the franchise(s) to a non-profit cor-
    poration? *No.* Did it consider
    municipal ownership? *No.*
20. Is there a provision in your franchise(s)
    for leased-channel access by commer-
    cial program suppliers? *Yes.*

Major Market Cable Survey
City of: Cincinnati

1. To whom was the cable franchise(s)
   awarded? *Warner Amex Cable Communications of Cinti., Inc.*
2. For how many years? *15.*
3. Total number of channels? *60.*
4. Does system have an interactive capacity? *Yes.*
5. How many tiers offered? *3.*
6. What is the projected cost of the
   system? *$58 million.*
7. How many homes will system pass?
   158,000.
8. What is the cost of basic cable service?
   $3.95/month.
9. How many channels offered in basic
   cable service? *24.*
10. What is the municipal fee? *5%.*
11. Percent of construction completed?
    90%.
12. Is construction on, ahead, or behind
    schedule? *Ahead.*
13. Date construction is scheduled for
    completion? *2/19/84.*
14. Number of public access channels? *15.*
15. Is there a dedicated fund for access
    programming? *Yes.*
16. Will candidates for public office be
    permitted to use public access channels
    as a matter of right? *—*
17. Have channels been reserved for
    municipal use? If so, how many? *1½.*
18. How many channels have been re-
    served for educational use? *3.*
19. Did your community consider award-
    ing the franchise(s) to a non-profit cor-
    poration? *No.* Did it consider
    municipal ownership? *Yes.*
20. Is there a provision in your franchise(s)
    for leased-channel access by commer-
    cial program suppliers? *Yes.*

Major Market Cable Survey
City of: Columbus, Ohio

1. To whom was the cable franchise(s) awarded? *Warner-Amex — QUBE.* *
2. For how many years? *10 — renewed 15.*
3. Total number of channels? *31.*
4. Does system have an interactive capacity? *Yes.*
5. How many tiers offered? *5.*
6. What is the projected cost of the system? *Unknown.*
7. How many homes will system pass? *91,898 Columbus; 140,609 Metro.*
8. What is the cost of basic cable service? *$11.95.*
9. How many channels offered in basic cable service? *12.*
10. What is the municipal fee? *3% of gross.*
11. Percent of construction completed? *99%.*
12. Is construction on, ahead, or behind schedule? *On.*
13. Date construction is scheduled for completion? *Oct. 1982.*
14. Number of public access channels? *3.*
15. Is there a dedicated fund for access programming? *Yes.*
16. Will candidates for public office be permitted to use public access channels as a matter of right? *Yes.*
17. Have channels been reserved for municipal use? If so, how many? *1.*
18. How many channels have been reserved for educational use? *1.*
19. Did your community consider awarding the franchise(s) to a non-profit corporation? *No.* Did it consider municipal ownership? *No.*
20. Is there a provision in your franchise(s) for leased-channel access by commercial program suppliers? *No.*

There are four cable operators serving Columbus.

Major Market Cable Survey
City of: Dayton, Ohio

1. To whom was the cable franchise(s) awarded? *Originally to Cypress, Southwest Cable. Sold to Viacom.*
2. For how many years? *15, extended by 2 years.*
3. Total number of channels? *35.*
4. Does system have an interactive capacity? *Yes.*
5. How many tiers offered? *Basic and 2 premium channels.*
6. What is the projected cost of the system? *—*
7. How many homes will system pass? *84,600 plus.*
8. What is the cost of basic cable service? *$9.95.*
9. How many channels offered in basic cable service? *29.*
10. What is the municipal fee? *5% of gross revenues.*
11. Percent of construction completed? *99%.*
12. Is construction on, ahead, or behind schedule? *Complete.*
13. Date construction is scheduled for completion? *9/81.*
14. Number of public access channels? *One active.*
15. Is there a dedicated fund for access programming? *Yes.*
16. Will candidates for public office be permitted to use public access channels as a matter of right? *Question decided by public access board.*
17. Have channels been reserved for municipal use? If so, how many? *1.*
18. How many channels have been reserved for educational use? *1.*
19. Did your community consider awarding the franchise(s) to a non-profit corporation? *No.* Did it consider municipal ownership? *No.*
20. Is there a provision in your franchise(s) for leased-channel access by commercial program suppliers? *Yes.*

Major Market Cable Survey
City of: Boston, Mass.

1. To whom was the cable franchise(s) awarded? *Cablevision of Boston Development Corp.*
2. For how many years? *15 years.*
3. Total number of channels? *208 channels (104 subscriber / 104 institutional).*
4. Does system have an interactive capacity? *Yes.*
5. How many tiers offered? *2 +.*
6. What is the projected cost of the system? *$110 million.*
7. How many homes will system pass? *250,000.*
8. What is the cost of basic cable service? *$2 for 52 channel basic service.*
9. How many channels offered in basic cable service? *52.*
10. What is the municipal fee? *3% of annual gross revenues.*
11. Percent of construction completed? *About 10%.*
12. Is construction on, ahead, or behind schedule? *On Schedule.*
13. Date construction is scheduled for completion? *Mid-1986.*
14. Number of public access channels? *On subscriber network: Initially 5 plus 1 municipal; up to 20 of total subscriber network capacity.*
15. Is there a dedicated fund for access programming? *Yes (5% of gross).*
16. Will candidates for public office be permitted to use public access channels as a matter of right? *Yes.*
17. Have channels been reserved for municipal use? If so, how many? *1 Subscriber channel; 10 Institutional channels.*
18. How many channels have been reserved for educational use? *As needed.*
19. Did your community consider awarding the franchise(s) to a non-profit corporation? *Yes.* Did it consider municipal ownership? *Yes.*
20. Is there a provision in your franchise(s) for leased-channel access by commercial program suppliers? *Yes.*

Major Market Cable Survey
City of: Grand Rapids, Michigan

1. To whom was the cable franchise(s)* awarded? *General Electric Cablevision Corp.*
2. For how many years? *10.*
3. Total number of channels? *30.*
4. Does system have an interactive capacity? *Yes.*
5. How many tiers offered? *Basic & family package & 3 individual pay movie channels.*
6. What is the projected cost of the system? *N/A.*
7. How many homes will system pass? *69,857.*
8. What is the cost of basic cable service? *$8.50 per month.*
9. How many channels offered in basic cable service? *24.*
10. What is the municipal fee? *3% of gross subscriber revenues.*
11. Percent of construction completed? *99.2%.*
12. Is construction on, ahead, or behind schedule? *Behind by 0.8%.*
13. Date construction is scheduled for completion? *Probably never 100%.*
14. Number of public access channels? *4.*
15. Is there a dedicated fund for access programming? *Yes.*
16. Will candidates for public office be permitted to use public access channels as a matter of right? *Yes.*
17. Have channels been reserved for municipal use? If so, how many? *1 channel.*
18. How many channels have been reserved for educational use? *2: 1 for K-12, 1 for Higher Ed.*
19. Did your community consider awarding the franchise(s) to a non-profit corporation? *N/A.* Did it consider municipal ownership? *No.*
20. Is there a provision in your franchise(s) for leased-channel access by commercial program suppliers? *Yes.*

Major Market Cable Survey
City of: Kansas City, Mo.

1. To whom was the cable franchise(s) awarded? *American Cablevision (subsidiary of ATC).*
2. For how many years? *15.*
3. Total number of channels? *36 (active & inactive).*
4. Does system have an interactive capacity? *Yes.* *
5. How many tiers offered? *Essentially 2. Basic service includes everything except 3 movie channels, which can be added as a group or separate.*
6. What is the projected cost of the system? *$30 million.*
7. How many homes will system pass? *140,000.*
8. What is the cost of basic cable service? *$10.50.*
9. How many channels offered in basic cable service? *28 (active).*
10. What is the municipal fee? *3% of gross revenue.*
11. *90 to 95%.*
12. Is construction on, ahead, or behind schedule? *A bit behind.*
13. Date construction is scheduled for completion? *Summer '83.*
14. Number of public access channels? *Five.*
15. Is there a dedicated fund for access programming? *No.*
16. Will candidates for public office be permitted to use public access channels as a matter of right? *Not as a right.*
17. Have channels been reserved for municipal use? If so, how many? *Yes. One.*
18. How many channels have been reserved for educational use? *One.*
19. Did your community consider awarding the franchise(s) to a non-profit corporation? *Yes.* ** Did it consider municipal ownership? *Yes, briefly.*
20. Is there a provision in your franchise(s) for leased-channel access by commercial program suppliers? *Yes.*

*Cables will handle 2-way communication but no hardware is in place to do this.
**But none came forward.

Major Market Cable Survey
City of: Albany, N.Y.

1. To whom was the cable franchise(s) awarded? *ATC (Capitol Cablevision).*
2. For how many years? *15 years with two 5 year renewal options.*
3. Total number of channels? *30 channels.*
4. Does system have an interactive capacity? *Yes — Two-way capable.*
5. How many tiers offered? *1 basic level with 2 premium channels.*
6. What is the projected cost of the system? *$2 million (approximately).*
7. How many homes will system pass? *43,000 homes.*
8. What is the cost of basic cable service? *$9.75 for first outlet; $4.00 for each additional.*
9. How many channels offered in basic cable service? *27 Channels.*
10. What is the municipal fee? *3% of gross revenues.*
11. Percent of construction completed? *100% as required by franchise.*
12. Is construction on, ahead, or behind schedule? *Completed.*
13. Date construction is scheduled for completion? *Completed.*
14. Number of public access channels? *One.*
15. Is there a dedicated fund for access programming? *No.*
16. Will candidates for public office be permitted to use public access channels as a matter of right? *Yes, based on right as resident.*
17. Have channels been reserved for municipal use? If so, how many? *Yes. One.*
18. How many channels have been reserved for educational use? *One.*
19. Did your community consider awarding the franchise(s) to a non-profit corporation? *Unknown.* Did it consider municipal ownership? *Unknown. Franchise awarded in 1968.*
20. Is there a provision in your franchise(s) for leased-channel access by commercial program suppliers? *No.*

Major Market Cable Survey
City of: Buffalo, New York

1. To whom was the cable franchise(s) awarded? *Cablescope Inc.*
2. For how many years? *15 years commencing in 1971.*
3. Total number of channels? *31 on two tiers plus premiums (10 on basic)*
4. Does system have an interactive capacity? *Not presently.*
5. How many tiers offered? *2.*
6. What is the projected cost of the system? *Not applicable.*
7. How many homes will system pass? *Currently 60,000 subscribers.*
8. What is the cost of basic cable service? *$8.50.*
9. How many channels offered in basic cable service? *10.*
10. What is the municipal fee? *5% franchise fee.*
11. Percent of construction completed? *100%.*
12. Is construction on, ahead, or behind schedule? *N/A.*
13. Date construction is scheduled for completion? *N/A.*
14. Number of public access channels? *1.*
15. Is there a dedicated fund for access programming? *No.*
16. Will candidates for public office be permitted to use public access channels as a matter of right? *?*
17. Have channels been reserved for municipal use? If so, how many? *1.*
18. How many channels have been reserved for educational use? *Currently in planning stages.*
19. Did your community consider awarding the franchise(s) to a non-profit corporation? *No.* Did it consider municipal ownership? *Currently under consideration.*
20. Is there a provision in your franchise(s) for leased-channel access by commercial program suppliers? *No.*

Major Market Cable Survey
City of: Oklahoma City, Oklahoma

1. To whom was the cable franchise(s) awarded? *Cox Cable.*
2. For how many years? *20.*
3. Total number of channels? *35.*
4. Does system have an interactive capacity? *No shadow cable.*
5. How many tiers offered? *Basic plus 4 pays.*
6. What is the projected cost of the system? *—*
7. How many homes will system pass? *155,000.*
8. What is the cost of basic cable service? *$10.00.*
9. How many channels offered in basic cable service? *31.*
10. What is the municipal fee? *3% of gross revenues.*
11. Percent of construction completed? *Substantially complete.*
12. Is construction on, ahead, or behind schedule? *On.*
13. Date construction is scheduled for completion? *2/83.*
14. Number of public access channels? *3.*
15. Is there a dedicated fund for access programming? *No.*
16. Will candidates for public office be permitted to use public access channels as a matter of right? *No.*
17. Have channels been reserved for municipal use? If so, how many? *1.*
18. How many channels have been reserved for educational use? *1.*
19. Did your community consider awarding the franchise(s) to a non-profit corporation? *No.* Did it consider municipal ownership? *No.*
20. Is there a provision in your franchise(s) for leased-channel access by commercial program suppliers? *Yes.*

Major Market Cable Survey
City of: Charleston, S.C.

1. To whom was the cable franchise(s) awarded? *Originally to Charleston Cable TV Co., now Storer Cable of Carolina, Inc.*
2. For how many years? *15.*
3. Total number of channels? *30.*
4. Does system have an interactive capacity? *No.*
5. How many tiers offered? *Basic only.*
6. What is the projected cost of the system? *?*
7. How many homes will system pass? *101,351.*
8. What is the cost of basic cable service? *$7.50.*
9. How many channels offered in basic cable service? *22.*
10. What is the municipal fee? *3%.*
11. Percent of construction completed? *100%.*
12. Is construction on, ahead, or behind schedule? *—*
13. Date construction is scheduled for completion? *January 1, 1982.*
14. Number of public access channels? *1.*
15. Is there a dedicated fund for access programming? *No.*
16. Will candidates for public office be permitted to use public access channels as a matter of right? *No.*
17. Have channels been reserved for municipal use? If so, how many? *One.*
18. How many channels have been reserved for educational use? *One.*
19. Did your community consider awarding the franchise(s) to a non-profit corporation? *No.* Did it consider municipal ownership? *No.*
20. Is there a provision in your franchise(s) for leased-channel access by commercial program suppliers? *No.*

Major Market Cable Survey
City of: Memphis, Tennessee

1. To whom was the cable franchise(s) awarded? *Memphis CATV, Inc.*
2. For how many years? *15 years.*
3. Total number of channels? *35.*
4. Does system have an interactive capacity? *Yes.*
5. How many tiers offered? *1.*
6. What is the projected cost of the system? *$42,000,000.00.*
7. How many homes will system pass? *250,000.*
8. What is the cost of basic cable service? *$8.25 per month.*
9. How many channels offered in basic cable service? *33.*
10. What is the municipal fee? *5%.*
11. Percent of construction completed? *100%.*
12. Is construction on, ahead, or behind schedule? *Ahead.*
13. Date construction is scheduled for completion? *September, 1982.*
14. Number of public access channels? *1.*
15. Is there a dedicated fund for access programming? *No.*
16. Will candidates for public office be permitted to use public access channels as a matter of right? *No.*
17. Have channels been reserved for municipal use? If so, how many? *1.*
18. How many channels have been reserved for educational use? *2.*
19. Did your community consider awarding the franchise(s) to a non-profit corporation? *No.* Did it consider municipal ownership? *No.*
20. Is there a provision in your franchise(s) for leased-channel access by commercial program suppliers? *Yes.*

Major Market Cable Survey
City of: Dallas, Texas

1. To whom was the cable franchise(s) awarded? *Warner Amex.*
2. For how many years? *15.*
3. Total number of channels? *80.*
4. Does system have an interactive capacity? *Yes.*
5. How many tiers offered? *3 Tiered Rates.*
6. What is the projected cost of the system? *$146,000,000.*
7. How many homes will system pass? *381,500.*
8. What is the cost of basic cable service? *$9.95 per mo. (3 Tiers).*
9. How many channels offered in basic cable service? *20 channels.*
10. What is the municipal fee? *5% street rental fee.*
11. Percent of construction completed? *approx. 50% — 4-18-83.*
12. Is construction on, ahead, or behind schedule? *probably ahead.*
13. Date construction is scheduled for completion? *October 1985.*
14. Number of public access channels? *One.*
15. Is there a dedicated fund for access programming? *No.*
16. Will candidates for public office be permitted to use public access channels as a matter of right? *No.*
17. Have channels been reserved for municipal use? If so, how many? *Two.*
18. How many channels have been reserved for educational use? *Seven.*
19. Did your community consider awarding the franchise(s) to a non-profit corporation? *No.* Did it consider municipal ownership? *Not extensively.*
20. Is there a provision in your franchise(s) for leased-channel access by commercial program suppliers? *Yes.*

Major Market Cable Survey
City of: Ft. Worth, Texas

1. To whom was the cable franchise(s)* awarded? *Sammons Communications.*
2. For how many years? *15.*
3. Total number of channels? *122.*
4. Does system have an interactive capacity? *Yes.*
5. How many tiers offered? *4.*
6. What is the projected cost of the system? *$85-91 million.*
7. How many homes will system pass? *175,000.*
8. What is the cost of basic cable service? *$3.95/5.95/7.95 (by tier).*
9. How many channels offered in basic cable service? *20/35/51.*
10. What is the municipal fee? *5% gross.*
11. Percent of construction completed? *35%.*
12. Is construction on, ahead, or behind schedule? *On.*
13. Date construction is scheduled for completion? *April, 1985.*
14. Number of public access channels? *10.*
15. Is there a dedicated fund for access programming? *Yes.*
16. Will candidates for public office be permitted to use public access channels as a matter of right? *Yes.*
17. Have channels been reserved for municipal use? If so, how many? *Yes — One.*
18. How many channels have been reserved for educational use? *Six.*
19. Did your community consider awarding the franchise(s) to a non-profit corporation? *Yes.* Did it consider municipal ownership? *Yes.*
20. Is there a provision in your franchise(s) for leased-channel access by commercial program suppliers? *Yes.*

Major Market Cable Survey
City of: Norfolk, Virginia

1. To whom was the cable franchise(s) awarded? *Cox Cable Communications, Inc.*
2. For how many years? *15 years.*
3. Total number of channels? *35.*
4. Does system have an interactive capacity? *Not yet operational.*
5. How many tiers offered? *Five.*
6. What is the projected cost of the system? *$15 million*
7. How many homes will system pass? *80,000.*
8. What is the cost of basic cable service? *$8.95.*
9. How many channels offered in basic cable service? *25.*
10. What is the municipal fee? *3% of total gross subscriber revenues.*
11. Percent of construction completed? *100%.*
12. Is construction on, ahead, or behind schedule? *On schedule.*
13. Date construction is scheduled for completion? *3/80.*
14. Number of public access channels? *One.*
15. Is there a dedicated fund for access programming? *No.*
16. Will candidates for public office be permitted to use public access channels as a matter of right? *No.*
17. Have channels been reserved for municipal use? If so, how many? *Yes; one.*
18. How many channels have been reserved for educational use? *One.*
19. Did your community consider awarding the franchise(s) to a non-profit corporation? *No.* Did it consider municipal ownership? *Not really.*
20. Is there a provision in your franchise(s) for leased-channel access by commercial program suppliers? *Yes.*

Major Market Cable Survey
City of: Tacoma, Washington

1. To whom was the cable franchise(s) awarded? *Teleprompter; now Group W Cable.*
2. For how many years? *25.*
3. Total number of channels? *29 possible.*
4. Does system have an interactive capacity? *No.*
5. How many tiers offered? *2 basic, pay.*
6. What is the projected cost of the system? *—*
7. How many homes will system pass? *App. 65,000.*
8. What is the cost of basic cable service? *$9.95.*
9. How many channels offered in basic cable service? *21.*
10. What is the municipal fee? *8% B&O Tax.*
11. Percent of construction completed? *80-85%.*
12. Is construction on, ahead, or behind schedule? *On.*
13. Date construction is scheduled for completion? *1984.*
14. Number of public access channels? *1.*
15. Is there a dedicated fund for access programming? *No.*
16. Will candidates for public office be permitted to use public access channels as a matter of right? *Only re: FCC rules.*
17. Have channels been reserved for municipal use? If so, how many? *4.*
18. How many channels have been reserved for educational use? *1.*
19. Did your community consider awarding the franchise(s) to a non-profit corporation? *No.* Did it consider municipal ownership? *No.*
20. Is there a provision in your franchise(s) for leased-channel access by commercial program suppliers? *Present now but not in franchise.*

Major Market Cable Survey
City of: Seattle, Washington

1. To whom was the cable franchise(s) awarded? *Group W.* *
2. For how many years? *15 years.*
3. Total number of channels? *35.*
4. Does system have an interactive capacity? *No.*
5. How many tiers offered? *4.*
6. What is the projected cost of the system? —
7. How many homes will system pass? —
8. What is the cost of basic cable service? *$10.55.*
9. How many channels offered in basic cable service? *17.*
10. What is the municipal fee? *6% Utility & Occupation Tax.*
11. Percent of construction completed? —
12. Is construction on, ahead, or behind schedule? *Ahead.*
13. Date construction is scheduled for completion? *1993.*
14. Number of public access channels? *3.*
15. Is there a dedicated fund for access programming? *No.*
16. Will candidates for public office be permitted to use public access channels as a matter of right? —
17. Have channels been reserved for municipal use? If so, how many? *1.*
18. How many channels have been reserved for educational use? *1.*
19. Did your community consider awarding the franchise(s) to a non-profit corporation? *No.* Did it consider municipal ownership? *No.*
20. Is there a provision in your franchise(s) for leased-channel access by commercial program suppliers? *No.*

Major Market Cable Survey
City of: Milwaukee

1. To whom was the cable franchise(s) awarded? *Warner Amex.*
2. For how many years? *15.*
3. Total number of channels? *108.*
4. Does system have an interactive capacity? *Yes.*
5. How many tiers offered? *2 + QUBE Service (Universal & Basic).*
6. What is the projected cost of the system? *$95 million*
7. How many homes will system pass? *253,000.*
8. What is the cost of basic cable service? *$4.95 per month.*
9. How many channels offered in basic cable service? *72.*
10. What is the municipal fee? *5%.*
11. Percent of construction completed? *0 (not yet begun).*
12. Is construction on, ahead, or behind schedule? *N.A.*
13. Date construction is scheduled for completion? *39 months.*
14. Number of public access channels? *13.*
15. Is there a dedicated fund for access programming? *Yes.*
16. Will candidates for public office be permitted to use public access channels as a matter of right? *No.*
17. Have channels been reserved for municipal use? If so, how many? *Yes — 2.*
18. How many channels have been reserved for educational use? *2.*
19. Did your community consider awarding the franchise(s) to a non-profit corporation? *Yes.* Did it consider municipal ownership? *Yes.*
20. Is there a provision in your franchise(s) for leased-channel access by commercial program suppliers? *No.*

Index